Indigenous Rights to the City

This book breaks new ground in understanding urban indigeneity in policy and planning practice. It is the first comprehensive and comparative study that foregrounds the complex interplay of multiple organisations involved in translating indigenous rights to the city in Latin America, focussing on the cities of La Paz and Quito.

The book establishes how planning for urban indigeneity looks in practice, even in seemingly progressive settings, such as Bolivia and Ecuador, where indigenous rights to the city are recognised within constitutions. It demonstrates that the translation of indigenous rights to the city is a process involving different actor groups operating within state institutions and indigenous communities, which often hold conflicting interests and needs. The book also establishes a set of theoretical, methodological, and practical foundations for envisaging how urban indigenous planning in Latin America and elsewhere should be understood, studied, and undertaken: As a process which embraces conflict and challenges power relations *within* indigenous communities and *between* these communities and the state.

This book will appeal to practitioners, researchers, and students working within the fields of urban planning, urban development, and indigenous rights.

Philipp Horn is a Lecturer in the Department of Urban Studies and Planning at the University of Sheffield, UK. His research interests centre around urban indigeneity; ethno-racial justice; participatory planning; and inclusive urban development in cities of the global South, with a regional focus on Latin America.

Routledge Studies in Urbanism and the City

This series offers a forum for original and innovative research that engages with key debates and concepts in the field. Titles within the series range from empirical investigations to theoretical engagements, offering international perspectives and multidisciplinary dialogues across the social sciences and humanities, from urban studies, planning, geography, geohumanities, sociology, politics, the arts, cultural studies, philosophy and literature.

For more information about this series, please visit: www.routledge.com/series/RSUC

Indigenous Rights to the City

Ethnicity and Urban Planning
in Bolivia and Ecuador

Philipp Horn

Routledge
Taylor & Francis Group

LONDON AND NEW YORK

First published 2019 by Routledge

2 Park Square, Milton Park, Abingdon, Oxfordshire OX14 4RN

52 Vanderbilt Avenue, New York, NY 10017

Routledge is an imprint of the Taylor & Francis Group, an informa business

First issued in paperback 2020

British Library Cataloguing-in-Publication Data
A catalogue record for this book is available from the British Library

Library of Congress Cataloging-in-Publication Data
A catalog record has been requested for this book

ISBN: 978-1-138-57358-1 (hbk)
ISBN: 978-0-367-66168-7 (pbk)

Typeset in Times New Roman
by codeMantra

Contents

Figures

Tables

Preface and acknowledgements

This book is the outcome of seven years of comparative research on the role of indigeneity in urban policy and planning practice. Rather than looking at abstract, ideal-type models of 'what could be done' to address urban indigeneity, it is the first comprehensive and comparative study that foregrounds the complex interplay of multiple actors involved in translating indigenous rights to the city in Latin America and, particularly, in the cities of La Paz (Bolivia) and Quito (Ecuador). The book is written for people interested in recent political developments in these two countries but also offers an empirical and practical contextualisation of concepts such as the right to the city and planning for diversity. It should therefore also be of interest to a broader audience of critical scholars, activists, and policymakers working on topics such as urban inequality, interculturalism, minority rights, and alternatives to urban development and planning in different settings in the global South and North.

The book brings together findings from three different research projects. This includes, first, doctoral research undertaken from 2011 until 2015, funded by the University of Manchester's Alumni Association, on the topic of constitutional changes and urban indigeneity in Bolivia and Ecuador. Second, the book draws on empirical findings from postdoctoral research (2016 until 2018) on links between discourses on group inequality, as addressed in the Sustainable Development Goals, and Ecuador's domestic development priorities on promoting indigenous rights and the right to the city. This was funded by the Open University's Innovation, Knowledge and Development Research Centre (IKD). Third, the book contains findings from a research project on peri-urban land conflicts and tensions between collective indigenous tenure and individual tenure rights in La Paz, Bolivia (2016 until 2018), which was funded through a Royal Geographical Society (RGS) Small Research Grant. I thank the University of Manchester's Alumni Association, IKD, and the RGS for funding these research projects.

It is impossible to name all the individuals and organisations who have contributed to this book: I want to express my deep gratitude and thanks to the different indigenous residents of Chasquipampa and Ovejuyo in La Paz, the students and teachers of the Chaquiñán College, and the representatives of different indigenous communes and migrant organisations in Quito. I would also like to thank all the indigenous leaders, politicians, policymakers, and planners who consented to be interviewed for this research. I have learnt a great deal from the perspectives they brought to rethinking urban indigeneity and how to study and address the topic of indigenous rights to the city. It is important to mention that most indigenous research participants approached in La Paz and Quito wanted to stay anonymous, and, for this reason, their names are omitted and replaced with pseudonyms throughout this book.

In La Paz, I would like to thank the team of SOS Children's Villages who first introduced me to the neighbourhoods of Chasquipampa and Ovejuyo. I also thank the team of the Urban Programme of the National Union of Institutions for Social Action Work (UNITAS) for their constant practical and intellectual support. In Quito, I am particularly grateful to Rodrigo Ugsha Cuyo, who invited me to teach in the Chaquiñán College and introduced me to his fellow comrades of the Tiguan Association of Carriers and Commercial Vendors Residing in Quito (AECT-Q).

I would also like to thank those individuals who have offered substantive ideas and advice on literature, engaged in discussions, or critically reviewed and commented on draft chapters. I owe particular debts to the following:

- In Bolivia: Xavier Albo, Walter Arteaga, Ann Chaplin, Ana Carola Farell Rodriguez, Carlos Mamani Condori, Pablo Mamani, Pedro Pachaguaya, Raul Prada, Carlos Revilla, Raul Rodriguez Arancibia, and Godofredo Sandoval.
- In Ecuador: Ileana Almeida, Fernando Garcia, Eduardo Kingman, Regine Mader, Angela Meentzen, Freddy Michel, Raul Moscoso, Jose Yanez de Posa, and Jeremy Rayner.
- In the UK: At the University of Manchester, I would like to thank my former PhD supervisors Caroline Moser and Alfredo Stein as well as Nicola Banks, Armando Barrientos, Tanja Bastia, Jessica Hope, Melanie Lombard, Matthew Thompson, and Peter Wade. I am also grateful for the feedback received from members of the Global Urban Writing Group. At the Open University, I thank my former colleagues Alex Borda Rodriguez, Sara de Jong, Lorenza Fontana, Jean Grugel, Giles Mohan, and Theo Papaionnou as well as members of the Engaged Scholarship group and the International Development and Inclusive Innovation research group. At the University of Sheffield, I thank my colleagues in the Department of Urban Studies and Planning and members of the Critical Friends reading group.

Throughout the last two years, I presented chapter drafts at a number of international conferences and policy events, including Habitat III in Quito (Ecuador), the annual meeting of the 2018 Development Studies Association in Manchester (UK), the annual meeting of the 2016 Royal Geography Society in London (UK), the 2016 World Planning Schools Congress in Rio de Janeiro (Brazil), and the conference 2018 'Urban Dialogues Socioeconomic restructuring, political-territorial reconfiguration and new representations of the urban Bolivia' in Santa Cruz (Bolivia). I am thankful for comments and feedback received during these conferences.

I am also very grateful to Ruth Anderson and Faye Leerink from Routledge for all their encouragement and support, and to Elizabeth Teague, whose editorial support significantly improved this manuscript.

Finally, and most importantly, I would like to thank my parents, Ulrike and Reiner, and my partner, Sally, for their unwavering love and support.

1 Introduction

From inhabitants of the forest to the concrete jungle

Indigeneity, urbanisation, and planning in Latin America

The research presented in this book focusses on the indigenous peoples[1] of Latin America, who inhabited the territories of this region in the centuries preceding colonisation by Spain. Focussing particularly on Bolivia and Ecuador, the book demonstrates how indigenous peoples are increasingly affected by urbanisation. In this context, it examines the everyday political struggles of urban indigenous residents as well as attempts by planners and policymakers to address indigenous rights in cities.

Indigenous peoples are often portrayed as living in isolated rural territories and pristine natural settings. They are also considered to represent 'guardians of the forest' who offer solutions for a more sustainable world.[2] Yet, throughout the world, indigenous peoples are being displaced, criminalised, and expelled from their rural territories and natural habitats for a variety of reasons, including the intensification of the extraction of natural resources, violent conflict, and climate change (Bebbington and Bebbington 2011; Coombes, Johnson and Howitt 2012; Hope 2016; Postero 2017). In this context, indigenous peoples increasingly reside in urban concrete jungles. According to a United Nations (UN) Habitat Report on 'Urban Indigenous Peoples and Migration,' 40 per cent of the world's indigenous population already lived in cities in 2010, and this number is likely to increase to more than 60 per cent by 2020 (UN Habitat 2010). There exist important regional and intraregional variations (see Table 1.1).

In Latin America, during the colonial and early postcolonial periods, the city was associated with a specific group of inhabitants – 'white' Spaniards or people of mixed blood who were granted citizenship rights (Platt 1982). The city was portrayed as a 'Western' and modern place in which economic progress occurred (Hardoy 1989). In contrast, the countryside was often understood to be an indigenous place, in which Latin America's native population led a backward peasant lifestyle and preserved their ancestral, non-Western traditions (Dussel 1993; Klor de Alva 1992; Walsh 2010). Since the second half of the 20th century, a variety of indigenous urbanisation trends can be observed. These range from natural growth amongst indigenous groups

Table 1.1 Indigenous urbanisation trends in different regions of the world

Region/Country examples	Percentage of indigenous peoples living in cities in 2010
Latin America	35
• Bolivia	42
• Chile	64
North America	NA
• United States of America	40
• Canada	54
Africa	NA
• Kenya	90
Asia	NA
• Tamil Nadu (India)	15
Australia	NA
• Australia	74
• New Zealand	84

Source: Elaborated by the author using data provided by UN Habitat (2010).

that have always resided in urban areas to rural to urban migration and the urbanisation of the countryside (see Chapter 2 for a detailed discussion).

Urbanisation and the move to the city did not automatically lead to improvements in indigenous peoples' living conditions. Instead, the lived reality of urban indigenous peoples can best be understood through the notion of 'coloniality of power,' which, according to Quijano (2000, 2007), refers to the prevalence of racial, social, political, economic, and cultural hierarchies established during the colonial conquest. Hence, despite being highly diverse in terms of place of origin and cultural background, urban indigenous peoples are disproportionately poorer than other urban residents (del Popolo, Oyarce and Ribotta 2009). Living predominantly at the margins of urban society (in slums, in informal and precarious housing, or more recently in social housing complexes), they are also more likely than other ethno-racial groups to be subjected to discrimination and exclusion from access to tenure rights, housing, and basic urban services.

This does not mean, however, that urban indigenous peoples are passive victims of exclusion, discrimination, and spatial segregation. Throughout the region, urban indigenous peoples increasingly engage in struggles for inclusion, recognition, and rights (Bengoa 2000). This is particularly evident in Bolivia and Ecuador, where urban indigenous peoples took to the streets during events such as the 2003 gas war in the cities of La Paz and El Alto, and urban indigenous uprisings in 2000 in Quito (Becker 2010; Bengoa 2000; Guss 2006; Lazar 2008; Revilla 2011; Zibechi 2010). As part of these insurgent urban uprisings, protestors demanded formal recognition of specific indigenous rights around self-governance and prior consultation as well as universal rights to shelter, tenure, and basic services. They also demanded to be involved in decision-making processes within the cities in which they live. It is this combination of demands which are referred to in this book as indigenous rights to the city. This terminology is chosen as the aforementioned demands articulated by urban indigenous residents closely resemble what French critical theorist and radical thinker Henri Lefebvre (1968, 1991) referred to as the right to the city. According to Lefebvre, the right to the city should not be conflated with a legal approach to urban development. It represents a cry and demand by historically marginalised groups to appropriate urban space according to their interests and needs; to participate in decisions concerning urban planning, design, and management; and to be at the core of urban life (see also Butler 2012; Harvey 2008, 2012; Marcuse 2009; Merrifield 2011; Purcell 2002).

Consequences of the urban uprisings in La Paz, El Alto, and Quito included the ouster of national governments and the election of new progressive left-leaning governments. These governments ratified new constitutions, in 2008 in Ecuador and in 2009 in Bolivia (CPE Bolivia 2009; CPE Ecuador 2008), which, for the first time in the history of Latin America and the world, recognise indigenous rights to the city as part of a new intercultural, plurinational, and decolonial development model which follows indigenous cosmovision on *Buen Vivir/ Vivir Bien* (in English: Living well). In a nutshell, these constitutions require national and local government authorities to address the specific interests and demands of indigenous peoples – independent of their rural or urban residence – in intercultural policies and planning interventions in sectors as diverse as education, urban development, healthcare, citizen participation, and housing. In this sense, then, the indigenous right to the city shifted from being a cry and demand raised by urban insurgents to a constitutional right which should be translated in policy and planning practice.

Previous research demonstrated that Bolivia's and Ecuador's new constitutions are characterised by contradictions, particularly in the ways they define and address indigenous rights in cities. For example, while both constitutions recognise indigeneity as a core element of their country's rural and urban development agendas, the literature claims that specific indigenous rights continue to be restricted to rural population groups (Colque 2009; Goldstein 2013; Kingman 2012). According to Goldstein (2013), this is particularly clear in the case of Bolivia's new constitution, which allocates specific indigenous rights only to 'authentic,' 'backward,' and 'rural' people, who were classified as 'indigenous original peasants.' While some studies have pointed out a set of tensions and contradictions within the constitutional texts, little research has so far investigated what the indigenous right to the city means to different actors, including not only policymakers and planners but also urban indigenous residents themselves, and how these actors incorporate indigenous rights to the city into urban policy and planning practice. This book seeks to answer these questions and address these knowledge gaps.

Framing, questions, and contribution of this book

This book considers urban planning and policy practice as the key battleground for indigenous peoples to claim their specific rights to the city. Scholars from a range of academic disciplines have already developed a variety of perspectives on urban planning's role in addressing indigenous peoples. One body of work considers planning knowledge and procedures as colonial practices which outlive colonialism and serve Western interests while contributing to the systematic dispossession and exclusion of indigenous peoples (Hardoy 1989; Myers 2003; Porter 2010). Others emphasise that indigenous communities are by no means passive victims of exclusion but active planners of their own lives who, through their everyday practices, strategic engagement with government authorities, and processes of violent and non-violent resistance, challenge, unsettle, and transform the planning system in such a way that their specific interests, demands, and rights-based claims are considered (Jojola 2008; Lazar 2008; Porter and Barry 2016). Recognising the problems of state-led planning procedures and the potential of indigenous community-led practice, another body of research develops a range of theoretical, methodological, and practical ideas on 'what could be done' to meaningfully engage indigenous peoples in the planning process. This includes, for example, Leonie Sandercock's (2003) work on multicultural cities and Libby Porter's and Janice Barry's (2016) work on planning for coexistence in settler colonial societies in the global North.

This book takes inspiration from the aforementioned scholarship on planning's role in addressing indigenous peoples. But it also points out its limitations. In particular, the book moves beyond a dualistic representation of state-led policy and planning practice as cause for exclusion and dispossession, and indigenous community-led practice as provider of emancipatory and transformative solutions. Such representations prevent us from understanding 'what is actually' done by different actors involved in urban policy and planning practice in countries such as Bolivia and Ecuador, where the urban indigenous population is highly diverse and represents government authorities and planners as well as ordinary urban residents of distinct socio-economic, political, and cultural backgrounds. To capture this diversity, the book develops an analytical framework which considers the translation of indigenous rights to the city as a process involving a multiplicity of officials operating in institutions of urban governance as well as a diverse group of indigenous residents. These people are considered social actors who define their practices not only in accordance with the norms and rules outlined in the constitutions but also in relation to themselves (Touraine 2000).

It is social actors involved in urban governance – in Bolivia and Ecuador this mainly refers to national and local government officials (Crabtree and Chaplin 2013; Grugel and Riggirozzi 2012) – who are considered to play a key role in defining the translation of constitutional rights. Drawing on the practice-centric literature on public policy and urban planning

(Bourdieu 1977; Flyvbjerg 2009; Moser 1993; Yiftachel 2006), this book demonstrates that government officials may not always follow constitutional guidelines. Rather, they may define their practices in such a way as to conform with their personal views on indigeneity and urban development. Furthermore, the practices of social actors involved in urban governance are considered to be shaped by the specific historical, political, and institutional environments in which they operate. It is also emphasised that social actors involved in urban governance may undertake practices in such a way that they conform to the agenda of their political party as well as to the demands of the interest groups they supposedly represent.

In addition to focussing on the practices of government officials, this book considers urban indigenous residents as planners of their own lives who play an active role in claiming and addressing their indigenous rights to the city and thereby contribute to the building of more inclusive and ethno-racially just urban societies. Hence, to identify the specific interests and demands of urban indigenous people as well as their understandings of indigeneity, the book deploys an asset-accumulation framework (Appadurai 2004; Moser 2009). Further, building on de Certeau's (1984) conceptual work on tactics, the book investigates indigenous peoples' own contestation and political negotiation practices to access a portfolio of resources and rights, and to influence decision-making in urban policy and planning from the bottom-up.

Drawing on the analytical framework outlined briefly above and in further detail in Chapter 3, this book critically examines the practices of national and local government officials, policymakers, urban planners, and ordinary indigenous communities in two major cities: La Paz, Bolivia, and Quito, Ecuador. By comparing the practices of these different actors, the book uncovers a set of conflicting realities of indigeneity as lived experience and as a policy and planning category. The following central points arise from this book: First, it is argued that, in everyday life, urban residents who self-identify as indigenous share that they want to move ahead economically. Both, in La Paz and Quito, indigenous residents generally consider their current governments' socio-economic development agendas, which focus on economic redistribution, as positively enhancing their economic well-being. Otherwise, though, urban indigenous residents express multiple and sometimes contradictory understandings of indigeneity, leading them to articulate different interests and rights-based claims. This makes it difficult for policymakers and planners to come up with one coherent political agenda on urban indigeneity.

Second, the book discovers other reasons why government officials, policymakers, and planners often fail to implement new constitutional rights on urban indigeneity: These include conflicting political priorities of government staff as well as ongoing views of the urban as an historically non-indigenous, 'white,' and modern place. The book finds that legislative reform represents only one initial step towards more inclusive and pro-indigenous urban policy and planning practice: A change in the attitudes and personal views of government authorities, policymakers, and planners is also required. This needs to be coupled with long-term structural changes that decolonise urban politics and society. It is demonstrated that such changes are already underway – at the level of national government through the reform of education laws and at the level of local government in La Paz and Quito, where a set of indigenous government officials are involved in the creation of new institutions and programmes which seek to mainstream the indigenous right to the city into citywide policy and planning practice.

Third, the book depicts a set of conflicting realities in the way different members of indigenous communities claim their indigenous right to the city. It is demonstrated that it is mainly indigenous elderly male leaders who engage with government authorities on behalf of their communities. Yet, in processes of political negotiations, these leaders rarely represent the members of their communities and mainly address their own interests. This speaks to a trend that indigenous communities are by no means homogenous but divided social entities characterised by uneven power relations and conflict. The book also reveals, though, that members of indigenous communities, especially youth and women, are increasingly problematising and confronting uneven power relations *within* their own communities and *between* communities and the state.

Notes on methodology

A qualitative and comparative case-study approach was deployed to develop an in-depth understanding of urban indigeneity and the complex claims and attempts to address indigenous rights to the city. La Paz and Quito are not selected as random cases from which generalisations for the wider Latin American region could be drawn (Flyvbjerg 2006). Rather, these two cities are chosen as illustrative cases because of their unique position as capital cities of those countries which – for the first time in Latin American history – recognise urban indigenous peoples within constitutions. La Paz's and Quito's status as seats of national government institutions facilitated access to multiple social actors involved in processes of translating indigenous rights to the city, including officials in national and local governments and ordinary urban indigenous residents and their relevant community-based organisations (for a detailed discussion, see Chapter 4).

The comparative rationale of the research is twofold, including within- and between-case comparisons. In terms of the former, the book focusses on multiple social actors and historic moments within each city in order to demonstrate internal complexities, contradictions, and diversities in the understanding of indigeneity and processes of translating indigenous rights (Flyvbjerg 2006; Yin 2003). For this reason, both the cities, and specific areas within each city, are studied in relation to their wider surroundings. For example, the book compares how, within both cities, different government officials – guided by distinct personal interests and political views; representing the interests of different individuals and groups; and holding different positions within institutions, such as municipal or national governments – translate indigenous rights into public policy and urban planning practices. The case-study approach also addresses different time-periods. Current legal understandings of indigeneity, as well as associated practices to translate indigenous rights, are studied in relation to past understandings and practices. In its within-case comparison, the book contrasts urban policy and planning interventions with the everyday interests, aspirations, and demands for resources and rights of ordinary urban indigenous residents. It also compares how different indigenous residents themselves understand indigeneity, articulate claims for the indigenous right to the city, and engage in processes of contestation and political negotiation in order to influence decision-making in urban policy and planning.

Building upon this within-case comparison, the book compares claims and attempts to implement the right to the city between the two case-study cities – La Paz and Quito. Unlike major research on indigeneity and indigenous rights in Bolivia, Ecuador, and the wider Latin American region, which has mainly deployed individualising, universalising, and encompassing comparisons, this book deploys a variation-finding comparison.[3] An individualising comparison was, for example, used in Andrew Canessa's (2012) eminent study of 'intimate indigeneities' in a Bolivian highland community – Wilja Kjarka. Canessa relied on Wilja Kjarka to contrast understanding of indigenous identity among members of this community with the way indigeneity is addressed elsewhere in Bolivia and in Latin America. Universalising comparisons were used by other eminent scholars interested in the expansion of indigenous rights across Latin America in the 1990s and early 2000s (Andolina, Laurie and Radcliffe 2009; Assies, Zoomers and Haar 2000; Sieder 2002; van Cott 2000). These studies convincingly showed that constitutional reforms and processes of expanding indigenous rights undertaken by different national governments in Latin America were influenced by very similar factors; for example, they responded to the demands of a transnational indigenous movement and incorporated international legislation on indigenous rights into domestic law. As a consequence, this literature shows that most Latin American countries introduced similar constitutional and legal reforms – described by van Cott (2000) as a 'multicultural model.' Using this 'multicultural model' as a starting point, other studies, including van Cott's (2008) own work, have relied on encompassing comparisons to show how various national governments in Latin America officially recognised international legislation on indigenous rights but, due to a variety of internal historical, institutional, and political factors, implemented these rights differently within their respective countries (Bengoa 2000; Marti i Puig 2010; Yashar 2005).

This book certainly draws on elements of the different comparative approaches outlined before. Following individualising comparisons, it recognises that understandings of indigeneity are likely to differ among members of different indigenous communities. Similar to studies that were guided by a universalising and encompassing approach, it acknowledges that both Bolivia and Ecuador recognise international legislation on indigenous rights, such as the International Labour Organisations Convention 169 and the United Nations Declaration on the Rights for Indigenous Peoples, but are likely to implement these rights differently. Unlike previous comparative research on indigeneity, however, this book mainly relies on a variation-finding comparative approach. By using this approach, it recognises that Bolivia's and Ecuador's governments not only responded to international pressure when addressing urban indigenous peoples within new constitutions. In fact, as has been demonstrated in recent non-comparative studies on Ecuador and Bolivia (Becker 2011; Schavelzon 2013), the contents of the new constitutions predominantly responded to internal demands raised during previous processes of popular resistance. Different internal demands were also articulated by a variety of social movements and political parties that were involved in constitutional assemblies. Differences in internal processes have rarely been captured in previous comparative research, which has predominantly studied the incorporation and implementation of indigenous rights at the national and local levels in relation to global and regional political trends. This book, therefore, contributes to existing comparative research by demonstrating that understandings of indigeneity and processes of translating indigenous rights to the city into public policies and urban planning practices vary *between* La Paz and Quito because of the unique domestic factors that shaped these processes *within* each city.

To compare different understandings of indigeneity and processes of claiming and addressing indigenous rights to the city within and between the two cities, the book draws on qualitative data which were collected as part of my doctoral research during 11 months of fieldwork in La Paz and Quito in 2012 and 2013. During this period, 92 interviews were conducted; these included interviews with authorities at the levels of national and local governments as well as indigenous leaders and ordinary indigenous residents. In addition, focus-group exercises were conducted, involving a total of 175 indigenous residents. Focus groups made use of participatory urban appraisal techniques such as listing and ranking of problems, causal flow diagrams, and institutional mapping, which made it possible to develop a more nuanced knowledge of the diverse rights-based claims, demands, and associated bottom-up interventions of ordinary urban indigenous peoples. Information from interviews and focus groups was complemented with archival and document analyses, participant observation, and analysis of secondary literature on urban indigeneity in Latin America and elsewhere. The book also draws on more recent empirical material gathered during follow-up research trips to Quito (April–May 2016) and La Paz

(October–December 2016). During these follow-up trips, fresh information on practical understandings of indigenous rights to the city was collected, including 40 additional semi-structured interviews, five further focus groups involving indigenous youth and women, additional archival research, and up-to-date policy-document analysis. It is important to mention that most indigenous research participants approached in La Paz and Quito wanted to stay anonymous, and, for this reason, their names are omitted and re-placed with pseudonyms throughout this book.

It is also important to reflect on some of the limitations of the research methodology. A first limitation of this methodology, generic to case-study approaches, relates to the potential to generalise the findings to other in-stitutions, indigenous communities, cities, or countries. Nevertheless, they provide useful illustrations of the complex, dynamic, and changing nature of indigeneity as a legal category, a lived experience, and a policy and planning category. Second, this research does not fully respond to growing demands by indigenous scholars and activists who advocate for a shift in research engagement with indigenous communities, emphasising a co-productive, consensus-based, and decolonial approach in which research is planned, designed, implemented, and disseminated with and for indigenous peoples (Smith 1999; Wilson 2009). The research process was designed to be as in-clusive and open as possible. For example, prior to commencing fieldwork research agreements were signed with the selected indigenous communities. After data analysis, findings were shared during community workshops and with citywide indigenous representatives. Yet, as a doctoral student and, later, as a postdoctoral researcher, the author still controlled many of the core features of this study, including community selection; questions asked; analysis of the data; and writing of final outputs, such as this book. This was the case for a variety of reasons, including, first, the relatively rigid design of UK doctoral programmes, which centre on the student as the central ac-tor in developing and implementing the research, and, second, the physical distance between the author, a UK-based researcher, and the indigenous communities studied in this book. The limitations of such a research ap-proach are acknowledged in this book. But it is, perhaps, also important to emphasise some of the advantages of conducting research which is not fully based on principles of co-production and especially consensus-building. To do this, it is best to quote at length a young indigenous research participant from Quito, Sara:

> You will never be able to write a book that will gain approval from everyone in the community as some might disagree with your interpre-tation on how things work here. This is especially the case for our lead-ers. They want to control everyone here and they also want to control your research. It is good that they don't accompany you when you do interviews. Because, if they would, we would not tell you what we re-ally think. You see, the leaders here don't represent us youngsters. They

would not want you to write this down. But please write it down for our sake. Look at us through your eyes and give us a fresh perspective. We can then deal with the results ourselves.

(Interview, 15 March 2013)

Understood like this, then, the researcher's role as outsider to a community can unravel phenomena, problems, and solutions to which perhaps not every community member would consent. In the case of this research, this indeed included the identification of uneven power relations within indigenous communities (see Chapters 5 and 7), something which benefitted older leaders but disadvantaged indigenous youths like Sara. Writing this down on paper should not be seen as an attempt to disrupt community relations and trigger conflict between different community members. Instead, in line with Sara's testimony the book simply intends to present the analysis of the author (an outsider), with the aim to stimulate critical debate among indigenous community members and other research participants as well as general readers interested in the topics discussed here.

Structure of the book

The book is organised into two parts and a concluding chapter. Part I positions the research within conceptual debates around urban indigeneity and planning, establishes the conceptual framing, and offers contextual background information on the selected case-study cities, La Paz and Quito. Chapter 2 focusses on the central theme of this book: Urban indigeneity. The chapter draws on a historical review of conceptual and policy debates on urban indigeneity, with specific emphasis on the Latin American region. It is argued that there is not one but multiple meanings of urban indigeneity. It is demonstrated that urban indigeneity is best understood as a dynamic category which is fluid and constantly changing over time and space. Further, people who self-identify as indigenous within cities tend to belong to different ethnic groups and peoples. They may also have a different class background, represent migrants who came to the city from the countryside, or members of indigenous communities whose territories have been absorbed by processes of urban expansion. Despite all these differences, the chapter finds that colonisation, dispossession, and exclusion – from specific rights, access to secure tenure, services, and housing – represent common characteristics describing the lived reality of most urban indigenous peoples. It is argued that the historical construction of indigeneity as a political and planning category associated with rurality and urban marginality represents a key cause for the ongoing exclusion of indigenous peoples in cities. At the same time, the chapter demonstrates that urban indigenous peoples by no means represent passive victims of exclusion. Instead, they are active planners of their own lives who claim their specific right to the city and lay the groundwork for more inclusive, emancipatory, and decolonial urban alternatives.

Recognising that planning represents one of the key battlegrounds for indigenous right to the city struggles, Chapter 3 fleshes out an analytical framework for the study of urban indigeneity as a policy and planning category. This framework captures 'what is actually done' to address urban indigeneity in a Southern context. The chapter conceptualises urban indigenous planning as a process involving multiple actor groups which follow different rationalities. Guided by a rationality of governing, one group of actors operates in institutions associated with urban governance and defines how urban indigeneity and associated indigenous rights to the city are actually incorporated into legislation, policy, and urban planning interventions. It is argued that, in addition to legal documents which define specific indigenous rights, the practices of these actors are shaped by their own personal views, the demands of the individuals and groups whom they represent, and the specific institutional and structural environments in which they operate. A second group of actors represents indigenous peoples who, guided by a rationality of survival, not only represent the target group of policy and planning interventions but are also active planners of their own lives.

Chapter 4 provides some contextual background on the case-study cities, La Paz and Quito. It demonstrates that these cities represent illustrative cases for a comparative study of the role of urban indigeneity in policy and planning, mainly because they are the major urban centres of those countries which – for the first time in global history – recognised indigenous rights to the city in their national constitutions. In addition to outlining the history of these cities, Chapter 4 offers a summary on recent patterns of ethnoracial diversification, provides an overview of the cities' highly diverse urban indigenous populations, and introduces the various national and local government authorities involved in urban policy and planning practice.

Part II of the book focusses on the case-study cities of La Paz and Quito, and offers an in-depth comparative analysis of the role of urban indigeneity as a category of lived experience and as a policy and planning category. Chapter 5 is the first of three chapters that discuss how different social actors involved in urban planning understand and practically address indigeneity and associated indigenous rights to the city. It begins by identifying what urban indigeneity means to various residents of La Paz and Quito who self-identify as indigenous. It also discusses the key interests, demands and rights-based claims of these residents. It is argued that indigenous residents in both cities share aspirations for economic improvement. The chapter also demonstrates that indigenous peoples mainly articulate specific interests and demands through claims for land. Land claims relate to interests in gaining access to a variety of cultural, financial, physical, social, natural/productive, or political resources. Chapter 5 also highlights a range of important intra-group differences in land claims. For example, it is demonstrated that in both cities elderly indigenous residents, whose territories were previously affected by urban expansion, seek to preserve what remains of their collectively owned land for agricultural purposes (that is, land as a collective and productive resource) and thereby mobilise around their collective rights to territory. Younger indigenous residents living in the same areas also mobilise for their collective territorial rights but with a somewhat different underlying rationale. They often want to sell their 'ancestral' land for a good price on the urban land market (that is, they view the land as a financial resource). In short, the chapter backs the overarching argument of this book that urban indigenous peoples raise multiple and conflicting interests, demands, and rights-based claims. It is perhaps for this reason that government authorities find it difficult to come up with one coherent policy and planning agenda on urban indigeneity.

Chapter 6 focusses in detail on how government authorities address urban indigeneity in policy and planning interventions. It is evident in both cities that political changes induced by the governments led by Presidents Morales and Correa enhanced urban indigenous peoples' power, with some now holding important positions in national and local governments. Economic redistribution policies have also enhanced the economic well-being of indigenous peoples in both cities. The chapter, however, also reveals a gap between progressive constitutional rhetoric on the indigenous right to the city, which emphasises addressing urban indigenous interests and needs in every policy sector, and actual urban policy and planning practice. Focussing on La Paz and Quito, the chapter demonstrates how national and local government authorities have difficulty translating constitutional principles on urban indigeneity into policies and planning interventions because they (1) lack clear legislative and operational guidelines, (2) hold a range of preconceived notions of cities as non-indigenous spaces, (3) struggle to address the conflicting and contradictory interests of different indigenous groups, and (4) follow different political priorities. In addition to identifying conflicting realities in urban policy and planning, the chapter examines 'best practice' examples of pro-indigenous interventions. For La Paz, it discusses the work of the national government's vice ministry on intercultural relations and the municipal intercultural unit which seeks to mainstream indigenous collective rights into the work of diverse policy sector units working on urban land management, housing, healthcare, culture, and education. For Quito, it explores the work of district administrations that address indigenous rights, interests, and needs through targeted interventions in healthcare and cultural events in those neighbourhoods in which most residents are of indigenous descent. Drawing on these findings, Chapter 6 concludes by stating that legislative reform represents only one initial step towards more inclusive and pro-indigenous urban policy and planning. A change in attitudes among government authorities as well as long-term institutional reform and decolonisation of urban politics and society is also required. It is argued that, to achieve such long-term changes, it would be best to identify those practices that work best in the context of specific cities and to subsequently strengthen and mainstream such interventions across different policy sectors. This requires paying close attention not only to 'best practices' of national and local governments but also to the everyday politics and planning efforts of ordinary urban indigenous communities.

Chapter 7 discusses the role of indigenous peoples as 'planners' of their own lives, who, by engaging in processes of self-help, political negotiation, or contestation, seek to address their specific demands and rights-based claims within the political environment that governs them. The findings presented here reveal that it is often not every 'ordinary' urban indigenous resident but mainly community leaders that play an influential role in processes of political negotiation and contestation. In processes of political negotiations, however, indigenous leaders do not automatically represent the interests and demands of all the members of their communities. Rather, they often use their powerful position to enrich themselves personally or to provide close friends with access to resources. This speaks against a view of urban indigenous communities as harmonious and reciprocal entities. Instead, it is argued that, as in most urban neighbourhoods elsewhere in the world, indigenous communities represent divided places characterised by their own internal hierarchies, conflicts of interest, and unequal power relationships. By focussing on the practices of an indigenous youth tribe in La Paz's Chasquipampa and Ovejuyo neighbourhoods, the chapter also provides illustrations of bottom-up practices which seek to transform uneven power relations within and outside indigenous communities. It is argued that such bottom-up practices can, in the long term, lead to the consolidation of a more inclusive urban politics which leaves no indigenous resident behind.

The concluding chapter offers a synthesis of the key findings from this book and outlines a set of ideas for envisioning more inclusive and pro-indigenous urban planning.

Notes

1 There is no generally accepted definition on what constitutes an indigenous people (for a detailed discussion see Assies 1994; Dyck 1985). A commonly cited definition, accepted by the United Nations Permanent Forum on Indigenous Issues, is the one by José Martínez Cobo (1987: 29):

> Indigenous communities, peoples, and nations are those that, having a historical continuity with pre-invasion and pre-colonial societies that developed on their territories, consider themselves distinct from other sectors of the societies now prevailing in those territories, or parts of them. They form at present non-dominant sectors of society and are determined to preserve, develop, and transmit to future generations their ancestral territories, and their ethnic identity, as the basis of their continued existence as peoples, in accordance with their own cultural patterns, social institutions and legal systems.

2 For an illustration, see a recent campaign by Purpose, the Mesoamerican Alliance for People and Forests, the Coordinator of Indigenous Organizations of the Amazon River Basin (COICA), and the Ford Foundation: www.purpose. com/case_studies/guardians-of-the-forest/.

3 I draw on Charles Tilly's (1984: 83–84) typology of comparative research, including:

- Individualising comparisons which 'contrast specific instances of a given phenomenon as a means of grasping the peculiarities of each case.'
- Universalising comparisons which 'establish that every instance of a phenomenon follows essentially the same rule.'
- Encompassing comparisons which situate 'different instances at various locations within the same system, on the way to explaining their characteristics as a function of their varying relations to the system as a whole.' Variation-finding comparisons which 'establish a principle of variation in the character and intensity of a phenomenon by examining systematic differences among instances.'

Part I

Concepts and context

2 The emergence of urban indigeneity and the indigenous right to the city

Introduction

This chapter sets out with a discussion of continuities and changes in the meaning of indigeneity, with a particular focus on the Latin American countries of Bolivia and Ecuador. It critically examines how, from what, by whom, and for what reasons indigeneity was addressed differently at distinct moments in history and in distinct spaces, paying particular attention to relations between indigeneity and urbanisation.

A central point raised in this chapter is that indigeneity increasingly shifted from being a static category introduced by the colonisers and associated with rurality, social exclusion, and non-Western traditions to a much more dynamic social category which describes people living in both rural and urban areas who are of distinct ethnic, economic, social, cultural, and political backgrounds. Paradoxically, though, such shifts are hardly considered in legislative discourses and policy and planning practices which tend to reproduce static, colonial, and rural understandings of indigeneity. It is demonstrated that for urban indigenous peoples this has always represented a basis for everyday resistance, decolonisation, and liberation from ethnocentric urban power relations. In Bolivia and Ecuador, such indigenous resistance struggles stimulated political changes that culminated in the ratification of new constitutions which recognise specific indigenous rights to the city. First, though, let us reflect on the history that precedes these changes. The starting point for such an analysis must be the colonial conquest of the Americas – the moment which first introduced indigeneity as a social category.

Spanish colonialism and the emergence of rural 'Indians'

In 1491 there were no Indians in the Americas; in 1492 there were tens of millions.

(Canessa 2008: 354)

As highlighted in the aforementioned quote by Andrew Canessa, before the arrival of the colonisers, the geopolitical territory which today represents

Latin America was composed of diverse societies. Cities were the political, economic, and cultural centres of these societies. In today's Bolivia, Tiuhuanaco was the urban administrative centre of pre-Inca civilisations living in the area of Lake Titicaca until approximately 1200 AD. Subsequently, the Incas controlled most of the Andean region, with Cuzco in today's Peru being the urban heart of the Inca Empire and cities such as Chukiyapu and Kitu – today's La Paz and Quito – representing important administrative centres (Hardoy 1973).

Latin America experienced a dramatic change with the arrival of the Spanish colonisers who ignored differences among local population groups with distinct ethnic, linguistic, political, and cultural backgrounds, and described all the region's native population as 'Indians.'[1] The colonial conquest led to the eradication (e.g. through massacres and outbreaks of diseases) of almost half of Latin America's native population of approximately 100 million within three decades (Quijano 2005). The colonisers also destroyed most precolonial cities. On the ruins of these ancient cities emerged new urban centres from which the colonisers controlled the rural hinterlands (Hardoy 1989). In today's Bolivia, Chukiyapu became the city of La Paz (Albo 2005; Guss 2006). In today's Ecuador, Kitu became Quito (Zaaijer 1991).

The colonial conquest led to the creation of indigeneity as a social category. The colonisers established a political system of ethno-racial stratification that followed the Spanish model of blood politics. Based upon phenotype, cultural, and linguistic backgrounds, the colonisers drew political, economic, and spatial distinctions between different population groups (Fenton 2003; Harris 1995; Wade 2010). They divided Latin American societies into two different states: A Spanish republic and an 'Indian' republic (Platt 1982). The Spanish republic granted citizenship status to 'white' Spaniards and *criollos* (Spaniards born in the colony) who were also recognised as independent producers of commodities entitled to engage in wage labour and commercial exchange. In contrast, the 'Indian' republic was maintained through a 'pact of reciprocity' whereby 'Indians' had to pay a tribute to the colonisers in order to maintain a plot of rural land. Otherwise, 'Indians' were denied citizenship status and often were not allowed to inhabit cities. They predominantly served under semi-feudal conditions as peasants or miners on *encomiendas*.[2] Hence, being part of the 'Indian' republic was associated with social exclusion:[3] It meant being set apart and locked out spatially, culturally, politically, and economically from other ethno-racial groups, such as whites, *criollos*, or black slaves.[4]

For the colonisers, the introduction of indigeneity served to create an inferior other – the 'Indian' – who could be exploited and used to serve the interests of a superior group, the 'white' Europeans or *criollos*. Furthermore, by introducing an ethnocentric system based on domination, dispossession, and exploitation, the colonisers also managed to gain control over global commercial traffic, with the latter leading to regional economic expansion at home in Europe. It is precisely on these grounds that Latin American scholars such as Enrique Dussel (1993), Walter Mignolo (2000), and Anibal Quijano (1993, 2000) argue that the notion of capitalist modernity and its dialectical other – tradition – arose as a consequence of ethno-racial 'blood politics' in Latin America and not, as argued by Karl Marx, as a result of the rise of the bourgeoisie as a new revolutionary class in Europe. In the Latin American colonial context, modernity was associated with a specific place – the city – inhabited by specific people – 'white' Spaniards and *criollos* – and characterised by Western culture and economic progress. Meanwhile, tradition was associated with another place – the countryside – inhabited by specific people – 'Indians' – and characterised by non-Western culture and backwardness.

Indigeneity is often represented as a static category and historical product of colonialism which operates according to schemes generated by colonialism throughout postcolonial history until the present (Engerman and Sokoloff 2000; Klor de Alva 1992; Mahoney 2003; Quijano 2000; Porter 2010; Salomon 1988). Peruvian sociologist Anibal Quijano (2000) considers this as an example of the 'coloniality of power,' which he defines as the continued presence of social, political, economic, and cultural hierarchies established by the colonisers. Quijano (2006) illustrates this point by showing how the abolition of the *encomienda* system after the fall of the colonial empire only led to the introduction of a new but rather similar scheme – the *hacienda* system,[5] which left 'Indians' trapped under semi-feudal working conditions. Other scholars such as Les Field (1994) and Oren Starn (1991), who belong to what is now known as the 'cultural survival school' or 'Andeanism,' seek less to trace postcolonial continuities around processes of ethno-racial stratification and exclusion but rather to demonstrate cultural continuities. Their research focusses on remote rural communities in the Latin American Andean and Amazonian regions, and highlights how, throughout postcolonial history, those described as 'Indians' by the colonisers preserved non-Western cultures and ancient economic and political principles that stand in contrast to modern life in the 'white' urban core.

Static accounts may be criticised for reproducing colonial understandings of indigeneity, for exoticising and romanticising the cultural practices of rural indigenous peoples, and for creating false binaries between rural tradition and urban modernity (Stavenhagen 1981). In fact, indigeneity represented a much more dynamic social category already during colonialism and in the early postcolonial period. For example, historical studies by Xavier Albo (2005) and Fausto Reinega (1970) highlight that indigenous peoples were never passive victims of exclusion but actively resisted oppression in the colonial and early postcolonial periods. Examples include not only the siege to La Paz, Bolivia, laid in the 1780s by indigenous leader Tupak Katari and his wife Bartolina Sisa but also the indigenous revolt led by Fernando Daquilema during the 1860s in Ecuador. Similar indigenous revolts occurred in other (post)colonial societies in Latin America and other regions of the world (see, for example, Porter 2010; Russel 2005). Writing on Latin American colonial cities, Jorge Hardoy (1989) and Richard Morse (1978) further show how indigenous peoples were never expelled or fully excluded from urban life. Instead, indigenous peoples continued to live, albeit in dire conditions, within the peripheries of colonial and early postcolonial cities. In fact, the construction of Latin America's colonial cities is an outcome of often forced indigenous labour. As part of this process, though, indigenous peoples subtly incorporated precolonial symbols into colonial urban architecture, design, and arts. This is perhaps most visible in the interior design of colonial churches in Cuzco and Lima, Peru, where statutes depict St Mary wearing a triangular dress, symbolising the Andean mountains and spiritual connections to mother earth (in Spanish: Pachamama) – a sacred figure in Inca religion.

The aforementioned examples represent early efforts to decolonise Latin American society. Quijano (2007: 177) defines decolonisation as liberation from European ethnocentric power relations 'organized as inequality, discrimination, exploitation and domination.' Decolonisation, whether through violent resistance struggles or through the subtle incorporation of indigenous precolonial symbols within colonial urban landscapes, represents the dialectical other and logical response towards colonialism and coloniality. The next sections will further unpack this dialectical interplay by focussing on indigenous urbanisation trends in more recent periods of Latin American history, namely modernisation (1900s–1970s), neo-liberal multiculturalism (1980s–2000s), and (post-)neo-liberalism (2000s–present).

Modernisation (1900s–1970s): towards urban coloniality

Until the early 20th century, Latin America's native population was mainly referred to as 'Indians' (Canessa 2007). This changed at the beginning of the 20th century, a period associated with modernisation and 'deindianisation' (Larson 2004). The Mexican Revolution of 1910–1920 represents an important reference point for these changes (Bengoa 2000). Guided by indigenism, the Mexican Revolution introduced a process of acculturation (that is, cultural assimilation). It revalued the status of indigenous peoples by emphasising and idealising the mixed biological and cultural heritages (in Spanish this is referred to as *mestizaje*) of all Mexicans who were now granted citizenship rights (Canessa 2006). Revolutionary forces were also strongly inspired by Marxism and the rise of communism in Eastern Europe. Mexican revolutionaries shifted their focus away from ethno-racial relations to class relations, putting emphasis on promoting the rights of a new peasant class (the former 'Indians') to own land. With these changes came the abolition of the 'Indian' tribute and the *hacienda* system (Albo 2005). Reforms resembling those in Mexico occurred throughout Latin America, including in Bolivia – after the Bolivian revolution in 1952, in Ecuador – as part of land reforms in 1964 and 1970, in Guatemala – after a popular revolution in 1954, in Nicaragua – during the Nicaraguan Revolution in 1979, and in Peru – as part of land reforms in the 1950s.

Let us focus on the countries which are studied in further detail in this book – Bolivia and Ecuador. Here, national governments granted citizenship rights to indigenous peoples, abolished the *hacienda* system, and promised to redistribute land to those who worked it – indigenous peoples who were now ascribed as peasants (Albo 1987; Barsky 1984). In practice, however, these reforms rarely led to an improvement in the living conditions of indigenous peoples – the new peasant class. In Bolivia, land redistributions remained restricted to indigenous peasants living in the country's highland departments (regional subdivisions) of La Paz, Cochabamba, Chuquisaca, and Potosi. Here, excessive land parcelling was undertaken, leaving indigenous peasant families with small plots of land (Dunkerley 2007). In contrast to Bolivia's highlands, redistributive policies were not introduced in the lowland departments of Beni, Pando, Santa Cruz, and Tarija, where Bolivia's government favoured the promotion of large industrial agro-enterprises. Similarly, Ecuador's government followed development recommendations manifested within US President J.F. Kennedy's 'Alliance for Progress' programme and prioritised the creation of large-scale agro-enterprises over land redistribution (Korovkin 1992; Scheman 1988). A result of the land reforms in both countries was that indigenous peoples either received a plot of land which was too small for household subsistence or, in the case of Bolivia's lowlands and Ecuador, often remained landless.[6]

Since indigenous peasants often could not live on the land redistributed to them, they had to engage in processes of split-migration, meaning that some members stayed in the countryside while others moved to cities in search of work (Albo *et al.* 1981). Rural-to-urban migration was also an indirect result of the agricultural reforms that freed indigenous peoples from semi-feudal conditions, granted them citizenship status, and permitted them to relocate to other places. Another reason for rural-to-urban migration, particularly in Bolivia, was the severe droughts that occurred in the late 1950s and which led to a further decline in agricultural productivity and resulted in famines (Lazar 2008). Furthermore, in lowland Bolivia and in Ecuador, rural-to-urban migration mainly occurred because rural indigenous peoples remained landless, leaving them with no other option than to search for economic opportunities elsewhere (Espin 2012; Torrico Foronda 2011; Zaaijer 1991).

While only five per cent of indigenous peoples lived in Bolivian cities at the beginning of the 20th century, this number increased to more than 20 per cent by the mid-1950s (Klein 2011). Similarly, in Ecuador approximately three per cent of indigenous peoples lived in cities in the 1930s but this number rose to more than 15 per cent in the 1960s (Kingman 2012). In Bolivia, highland indigenous groups, often of Aymara and Quechua backgrounds, mainly migrated to cities such as La Paz, Cochabamba, Sucre, and Potosi. Meanwhile, lowland indigenous groups, representing among others Ayoreo, Chiquitano, Guarani, and Guarayo peoples, mainly migrated to cities such as Santa Cruz. Indigenous migration in Ecuador focussed mainly on larger cities such as Quito, Guayaquil, and Cuenca. Similar trends were observed in other Latin American countries such as Mexico and Peru where indigenous peoples mainly migrated to the capital cities of Mexico City and Lima (Matos Mar 1957; Oehmichen 2001).

The indigenous move towards the city led to an unprecedented rise of new low-income settlements in the urban periphery of major Latin American cities. This was particularly evident in Lima, Peru where urban indigenous migrants initially settled in densely populated colonial houses in the peripheries of the colonial centre and, at later stages, started occupying land and self-constructed houses in the growing unplanned urban peripheries – often referred to as *barriadas* (Matos Mar 1957). Self-construction, according to John F.C. Turner (1968, 1978), was as much a desperate act by indigenous migrants and other low-income groups to access housing in an environment where other options were simply too costly as it was a process of empowerment since people could take charge of the design, construction, and management of their own homes in a cost-efficient way. Similar trends around self-help housing initiatives in the urban periphery can be observed for first-generation urban indigenous migrants arriving to Bolivian cities such as La Paz and Ecuadorian cities such as Quito (see also Chapter 4).

Urbanisation also led to processes of indigenous identity transformation. Guided by modernisation and assimilation theories, Pierre van den Berghe (1974: 9), for example, argued that characteristics associated with indigeneity such as non-Western tradition or exclusion lose their hold in the city:

> The closer one comes to the larger urban centres and their interconnecting main roads and railways, the more processes of cultural hispanisation and the extension of bilingualism tend to blur ethnic distinctions and give more salience to class difference.

Findings from Bolivian and Ecuadorian cities challenge such modernist and assimilationist assumptions and reveal ongoing processes of othering by 'white' residents who treat indigenous migrants as 'Indians,' thereby reproducing patterns associated with coloniality within urban settings (Albo *et al.* 1981; Lazar 2008; Weismantel 2001). That does not mean, however, that urban indigenous migrants were passive victims of exclusion and discrimination. In fact, they often developed their own distinct urban ethnic identities – referred to, for example, as *cholos* (people of mixed race) and *mestizos* (people of mixed Spanish and indigenous descent) in Bolivia, Ecuador, and Peru or *ladinos* (Spanish-speaking 'white' people) in Guatemala. For example, urban indigenous *mestizos* in Cuzco, Peru strategically fused organisational modalities associated with traditional rural *ayllus*[7] and modern cities (de la Cadena 2000). In La Paz, Bolivia, urban indigenous migrant women created their own distinct *chola* identity which is perhaps visible in their particular clothing style – a wide skirt, the Manila shawl, and the Borsalino hat (Rivera

Cusicanqui 2010). Urban indigenous migrants also often retained strong ties to their rural communities, preserved their indigenous languages and traditions when interacting with each other but spoke in Spanish with urban residents of other ethno-racial backgrounds (Albo *et al.* 1981, 1983).

In short, modernisation led to profound processes of rural-urban restructuring with an increasing number of indigenous peoples now living in cities. Despite important legislative reforms, within cities indigenous peoples continued to be treated as second-class citizens and remained excluded from socio-economic opportunities available to urban 'whites.' In other words, ethno-spatial divisions *between* rural indigenous and 'white' urban areas established by the colonisers now materialised strongly *within* cities, leading to a situation of urban coloniality. At the same time, though, everyday efforts for decolonisation – whether expressed through the adaptation of distinct ethnic identities or the introduction of specific indigenous governance schemes within cities – also prevailed within urban settings. The next section illustrates how neo-liberal reforms accelerated processes of indigenous urbanisation without per se altering conditions of urban coloniality, leading to a further rise in indigenous resistance and decolonisation efforts.

Neo-liberal multiculturalism (1980s–2000s): rural rights and urban outlaws

Between the late 1970s and early 2000s the 'Indian' returned to Latin America and other postcolonial societies across the world, this time as a self-identifying indigenous person (Albo 1991; Assies 1994; Quijano 2006). The revitalisation of ethnic identity could mainly be observed in rural settings where, in a context of economic crisis and neo-liberal reforms, indigenous peoples, often with support from non-governmental organisations (NGOs) and the church, formed indigenous movements, started raising specific ethnic claims, and demanded specific rights from international organisations but also from their national governments (Andolina *et al.* 2009; Assies 2010; Marti i Puig 2010). Parallel to these processes in the countryside, another phenomenon occurred: The revitalisation of indigenous identities and the formation of indigenous organisations within cities – places where indigenous peoples remained outlawed from specific indigenous rights (Goldstein 2004; Lazar 2008; Kingman 2012). Let us reflect on these parallel yet interrelated rural and urban development processes with specific focus on Bolivia and Ecuador.

The emergence of rural indigenous movements and indigenous rights-based agendas

In Bolivia and Ecuador, two slightly different types of rural indigenous movement emerged since the 1970s. First, there was the rise of indigenous peasant movements – the Unified Syndical Confederation of Rural

Workers-Tupac Katari (CSUTCB-TK) founded in 1979 in Bolivia, and the National Federation of Indigenous Peasant Organisations (FENOC-I) founded in 1985 in Ecuador (Marti i Puig 2010). These movements united indigenous peasants who mobilised for land redistribution. This occurred in a context where Bolivia's and Ecuador's national governments, confronted by economic crisis and state bankruptcy, departed evermore from the promises of land reform for the peasants that they had made in previous decades. As a consequence, rural population groups became increasingly impoverished and disillusioned with the state. This led to an identity crisis among indigenous peoples whose peasant class identity was relatively new and unsettled. Questioning their class identity in a context of economic hardship, indigenous people often revitalised their ethnic identity (Albo 1991; Canessa 2006).

A second type of rural indigenous movement emerged with support from the church, local anthropologists, or NGOs who assisted indigenous groups to formulate a political agenda that moved beyond a focus on land redistribution (Andolina *et al.* 2009; Assies 2010; Bebbington 2007; Marti i Puig 2010). These movements raised specific rights-based claims around indigenous justice, the recognition of native languages, and intercultural education (Klein 2011; Yashar 2005). In Bolivia, such second-type indigenous movements include the Confederation of Indigenous Peoples of Bolivia (CIDOB) which was founded in 1982 by Amazonian indigenous groups as well as the National Council of Ayllus and Markas of Qullasuyu (CONAMAQ) which was founded in 1997 by highland indigenous groups (Klein 2011). In Ecuador, an example of a second-type indigenous movement would be the Confederation of Indigenous Movements of Ecuador (CONAIE) which was founded in 1986 and unites lowland and highland indigenous peoples (Yashar 2005).

These rural indigenous movements put increasing pressure on national governments[8] but also approached international organisations, which, from the late 1980s onwards, increasingly followed a rights-based approach to development.[9] Confronted by pressure from indigenous movements, international organisations such as the United Nations and the International Labour Organisation (ILO) held a series of international summits and conferences on human rights as well as on the social, economic, cultural, and specific-group rights of indigenous peoples in the late 1980s and 1990s. Outcome documents of these summits were a set of declarations which recognise indigenous peoples' rights. These include, for example, the 1989 ILO 169 Convention on Indigenous and Tribal Peoples, which prohibited all forms of discrimination against indigenous peoples and called for the strengthening of indigenous peoples' territorial, economic, social, and cultural rights, and for an end to ethno-racial inequalities (Anaya 2004).

Responding to internal pressure by indigenous movements and to new international legislation on indigenous rights, national governments throughout the Latin American region addressed indigenous rights through legislative reforms which Donna Lee van Cott (2000) refers to as the 'multicultural model' towards indigenous development. For example, Bolivia's and Ecuador's national governments introduced constitutional reforms, in 1993 and 1998, respectively, which acknowledged the ILO 169 Convention, recognised indigenous peoples as distinct groups with specific languages and traditions, granted indigenous peoples the right to manage and own land collectively, and guaranteed bilingual education in indigenous territories. The national governments subsequently introduced new state institutions – the Ministry of Indigenous Affairs (MAIPO) in Bolivia and the Council for the Development of Indigenous Nations and Peoples (CODENPE) and the National Directorate of Intercultural Bilingual Education (DINEIB) in Ecuador – that were responsible for implementing the constitutional rights of indigenous peoples (van Cott 2008). The national governments also indirectly strengthened the political weight of indigenous peoples through the introduction of decentralisation reforms, namely the 1994 Law of Popular Participation (LPP) in Bolivia and the 1997 Law of Decentralisation (LD) in Ecuador (Kohl and Farthing 2006; van Cott 2008). These laws expanded the competencies of decentralised municipalities, allocated a proportion of national tax revenues to municipal governments, and introduced direct elections at the municipal level. Bolivia's LPP also introduced the tool of participatory budgeting whereby municipal authorities should plan small-scale infrastructure projects together with members of indigenous peasant unions in the countryside and neighbourhood associations (so called *juntas de vecinos*–JJVVs) in cities (Kohl and Farthing 2006).

This new 'multicultural' indigenous development model had its limitations, particularly when it came to its implementation. A crucial factor which limited the implementation of indigenous rights throughout the Latin American region, including in Bolivia and Ecuador, was economic crisis and state bankruptcy. Governments responded to this problem by following advice from the World Bank and the International Monetary Fund and implementing an economic reform package which followed neo-liberal principles. This meant that governments devalued their currencies, freed exchange rates from state control, reduced public spending, eliminated price and wage controls, and encouraged large-scale private-sector investment in core industries such as mining and extraction of natural resources (Andolina *et al.* 2009; Postero 2007). A consequence of these reforms was that government authorities cut human and financial resources to fund relevant programmes or projects for indigenous peoples and prioritised capitalist interests over indigenous territorial rights. These reforms also led to widespread wage cuts and public – as well as private-sector – staff downsizing, induced through, for example, 'voluntary' withdrawal agreements in crucial sectors, such as mining.[10] Conditions in the agricultural sector also worsened. In a context of economic crisis, export demands declined as did domestic prices and demand, a situation that was compounded by declining real incomes. All this led to the effective dismantling of the agricultural sector, and especially of small agricultural units managed by indigenous peasants. At the same time, government authorities increasingly granted transnational companies permission to use local territories for purposes of commercial deforestation or resource extraction. Such activities often took place on territories inhabited by indigenous peoples who were now confronted by threats of expropriation (Andolina *et al.* 2009; Postero 2007).

In short, rural indigenous peoples were granted recognition of specific indigenous rights through the 'multicultural model' but access to these rights was restricted by the neo-liberal project. Charles Hale (2002) considers this to be the paradox of what he defines as 'neoliberal multiculturalism,' a project that endorses indigenous rights so that indigenous peoples can advance their own political agenda – as long as this does not disrupt the dominant capitalist order or, as I outline in the next section, interfere with previously established ethno-spatial patterns of structuring society into rural indigenous spaces and urban non-indigenous spaces.

Accelerated indigenous urbanisation and the absence of indigenous rights to the city

An important result of the aforementioned socio-economic changes was accelerated migration by indigenous peoples towards cities. By the late 1990s, more than a third of Bolivia's and around a quarter of Ecuador's indigenous peoples already lived in cities (Bengoa 2000). Within cities, however, indigenous peoples continued to be denied access to their newly established indigenous rights. A common explanation for this phenomenon is that the implementation of indigenous rights was restricted to rural areas, mainly because that is where mobilisation for specific indigenous rights-based agendas initially emerged but also because government authorities and representatives from donor organisations reproduced colonial understandings of indigeneity as a static category associated with tradition, backwardness, and rurality but not with modern urban life (Speiser 2004). Specific indigenous rights and development programmes, such as bilingual education, were therefore implemented predominantly in villages but very rarely in cities (Albo 2005; Hornberger 2000; Kingman 2012). Decentralisation reforms also mainly recognised the issue of indigeneity in rural but not in urban areas. This was particularly visible in Bolivia's LPP which reduced the involvement of indigenous peasant unions in participatory budgeting processes to rural areas. For cities, the LPP restricted participation rights to *vecinos* – a term historically associated with 'white' urban citizens. In other words, the LPP reproduced patterns of ethno-racial stratification; political exclusion; and, hence, urban coloniality.

In addition to being denied specific indigenous rights, urban indigenous peoples also confronted distinct problems in cities – such as racism, discrimination, unemployment, and missing basic services (Rivera Cusicanqui 2010). There now exists a rich repertoire of ethnographies on cities in Bolivia

(Gill 2000; Goldstein 2004, 2013; Lazar 2008; Postero 2007), Chile (Imilan 2010; Warren 2017), Ecuador (Colloredo-Mansfeld 2009; Kingman 2012; Swanson 2007), and Mexico (Oehmichen 2001) which highlights the precarious working conditions of indigenous peoples who are often confined to work in the informal sector as market vendors, food carriers, artisans, builders, or domestic workers. While a smaller number of indigenous peoples manage to make a decent income and enter the urban middle and upper classes – something that has been noted particularly for the emerging Aymara *bourgeoisie* in the Bolivian cities of La Paz and El Alto (Tassi 2010; Tassi *et al.* 2013) – the majority often earn just enough to provide for their basic needs and remain trapped in a situation of poverty (del Popolo *et al.* 2009). The socio-economic situation of urban indigenous peoples further worsened when municipal governments introduced neo-liberal reform policies and privatised core public services such as water and gas (Assies 2003; Perreault 2006).

In this difficult situation, urban indigenous peoples increasingly revitalised and mobilised around their ethnic identities. They seemingly questioned their belonging to the urban working class in a context of urban hardship, absence of government support, and ongoing discrimination directed towards them. Trends around ethnic mobilisation by urban indigenous peoples were particularly noteworthy in Bolivia where urban indigenous peoples started engaging in their own neighbourhood organisations in which they reproduced indigenous governance principles such as leadership rotation or collective work schemes (Albo 2006; Lazar 2008; Zibechi 2010). In the cities of Cochabamba and El Alto, indigenous peoples also started to rely on principles of indigenous community justice in order to cope with urban insecurity in the context of an absence of police and state forces (Goldstein 2004; Risor 2010). In Quito, indigenous groups began to form homeland associations which served as meeting points for indigenous migrants (Rojo 2012). Within these associations, indigenous peoples often organised festivals which helped to revive rural traditions and dances within the city. The preservation of indigenous traditions and culture has also been reported by David Guss (2006) who studied folkloric events such as the *Fiesta de Gran Poder* in La Paz, Bolivia. During this annual festival, urban indigenous folkloric associations marched through La Paz's city centre and performed traditional indigenous dances. Such performances not only helped indigenous peoples to revitalise their identity, they represented a decolonial act – the indigenous takeover of a historically 'white' city (Guss 2006). In the meantime, urban indigenous peoples also retained strong ties to their rural communities of origin where, by now, they could observe the increasingly important role placed by indigenous movements in negotiating and contesting for specific rights. Inspired by the work of rural indigenous movements, urban indigenous peoples started voicing similar demands for recognition of specific rights and political autonomy within the cities in which they lived (Albo 2006; Harris 1995; Lazar 2008).

Through engaging in urban collective action and the formation of community-based organisations (CBOs), indigenous peoples increasingly developed their own identity and also a political voice within the city. This became evident from the early 2000s onwards when urban indigenous peoples – in alliance with other low-income urban residents, rural peasants, miners, and rural indigenous movements – formed part of insurgent uprisings such as the 2000 Water War in Cochabamba, Bolivia (Assies 2003), the 2003 Gas War in La Paz/El Alto, Bolivia (Lazar 2008; Perreault 2006; Revilla 2011), or large-scale protests occurring in Quito, Ecuador in the early 2000s (Becker 2011; Bowen 2011). During these uprisings, indigenous peoples raised demands for what is here referred to as the indigenous right to the city – a cry and demand to be recognised as urban indigenous residents with distinct interests, and to be involved in urban decision-making processes. They protested against neo-liberal reform policies and processes of discrimination, inequality, exploitation, and 'white' domination, claiming their right to be recognised as urban indigenous residents with distinct interests and needs and to be involved in urban decision-making processes.

(Post-)neo-liberalism (2000s–present): incorporating the indigenous right to the city into constitutions

A consequence of these protests was the ousting of pro-neo-liberal governments in Bolivia (2003) and Ecuador (2005), and the emergence of new governments – led by President Evo Morales and the party Movement Towards Socialism (MAS) in Bolivia, and by President Rafael Correa and the party PAIS Alliance (AP) in Ecuador. Elected chiefly by the poor and the dispossessed indigenous majority, these left-leaning governments promised to introduce a 'post-neoliberal' agenda focussed on reasserting national control of macroeconomic policies and enhancing social welfare (Grugel and Riggirozzi 2012). For the first time in Latin American history, these governments also started seriously to consider the demands of urban indigenous residents. To incorporate the different demands of their electorate, the new governments convened constitutional assemblies. Bolivia's constitutional assembly was composed not only of all political parties represented in parliament but also of civil society representatives. Social movements and civil society organisations involved in the constitutional assembly included indigenous movements such as CSUTCB-TK, CIDOB, and CONAMAQ; labour unions representing miners and coca workers; urban neighbourhood associations and social movements representing the country's ethno-racially diverse urban population; and regional political groups representing lowland departments, such as Santa Cruz, Pando, Beni, and Tarija (Postero 2013; Schavelzon 2013). Unlike in Bolivia, Ecuador's constitutional assembly did not involve social movements but included only elected political parties (Becker 2011, 2013).

The new constitutions are based on the alternative developmental models of *Vivir Bien* (Bolivia) and *Buen Vivir* (Ecuador). *Vivir Bien/Buen Vivir* originates in indigenous worldviews; it emphasises that humans and nature should coexist in harmony and that collective interests should be prioritised over individual needs (Gudynas 2011). Following such an understanding, no one can live well if others live badly. To ensure the good life of all Bolivian and Ecuadorians, the constitutions outline a vast list of universal rights and political promises, guaranteeing citizens the rights to enjoy nature, universal healthcare, education, pensions, sports, and free time. Focussing specifically on urban settings, both constitutions guarantee citizens universal rights to housing, water, sanitation, public spaces, and participation. Ecuador's constitution introduces the principle of the 'right to the city' whereby every urban resident is 'entitled to the full enjoyment of the city and its public spaces, under the principles of sustainability, social justice, respect for different urban cultures and balance between urban and rural areas' (CPE Ecuador 2008, Article 31).

Both constitutions recognise the intercultural character of their societies and the rights of specific groups – particularly of indigenous peoples but also of other marginalised groups, such as women, Afrodescendants, people with disabilities, and the elderly. The constitutions further enhance previously established indigenous rights. Both Bolivia's (CPE Bolivia 2009, Article 30.1) and Ecuador's constitutions (CPE Ecuador 2008, Article 57) emphasise a set of specific indigenous rights including the right to collective land titles, the right of indigenous peoples to freely practise their culture and traditions, the right to autonomous management and governance of their territories, the right to prior consultation about interventions taking place on their territories, the right to self-government, and the right to exercise justice according to their own communitarian principles. Ecuador's constitution emphasises that indigenous rights apply within the places in which indigenous people live; no clear rural or urban specification is made. By contrast, Bolivia's constitution restricts specific indigenous rights to so-called indigenous original peasants (IOPs). By defining indigenous peoples through the IOP category, the members of Bolivia's constitutional assembly attempted to integrate different rural indigenous groups – indigenous peasants assembled in the CSUTCB-TK as well as indigenous groups assembled in the CIDOB or CONAMAQ – which mobilise around their status as 'authentic' or original indigenous peoples (Fontana 2014). Authors such as Colque (2009) and Goldstein (2013) have, however, argued that the focus on these different rural groups meant that the rights of urban indigenous peoples were not adequately addressed and that, as a result, urban-dwellers remained outlawed from specific indigenous rights.

In other sections, however, the Bolivian constitution addresses the topic of urban indigeneity. Article 218, for example, recognises cities as urban intercultural communities composed of indigenous peoples and other ethno-racial groups. These groups have the right to inhabit urban space according to their own interests and needs, and the latter should be ac-knowledged in sector policies such as intercultural education (Article 17), intercultural healthcare (Article 18), housing (Article 19), provision of basic public services (Article 20), and land management (Article 393). Further-more, Bolivia's constitution establishes that the participation of the coun-try's diverse population is central to the democratic management of cities (Article 241). Citizen participation schemes should not, therefore, discrim-inate according to gender or ethnicity but should involve different social groups in the design, evaluation, and implementation of policy and plan-ning interventions (Articles 26, 241).

The rights outlined in Bolivia's and Ecuador's constitutions closely re-semble what French critical theorist Henri Lefebvre (1968) would refer to as the right to the city[11] – a right to appropriate urban space according to ordinary people's interests and needs; a right to participate in decisions con-cerning urban planning, design, and management; and a right associated with centrality and being at the core of urban life. Bolivia's and Ecuador's constitutions not only recognise these components – in fact, Ecuador's con-stitution explicitly mentions the right to the city – but also emphasise their application for urban indigenous peoples. Both constitutions recognise that indigenous people's specific interests and needs should be considered in decisions around the making of urban space, but they also emphasise that indigenous organising modalities should be respected in participatory pro-cesses. It is these particular elements of the constitution that are referred to in this book as the indigenous right to the city.

Until now, little research has been conducted into the implementation of the new constitutionals. The few studies that have investigated advances and ongoing problems in the implementation of specific indigenous rights have tended to focus on rural areas (Escobar 2010; Filho and Goncalvez 2010; McNeish 2013; Postero 2017; Tockman and Cameron 2014; Walsh 2011). On the one hand, these studies note the tremendous positive effects the consti-tutional reforms and associated implementation practices have on revaluing indigenous peoples' identity and status in both countries. Racism has been on a decline, mainly due to anti-racism laws implemented by the govern-ments. Yet, as was the case in the neo-liberal period, indigenous rights con-tinue to be subordinated to principles of economic development. In both Bolivia and Ecuador, the national governments decided to sell rights over land held by indigenous communities to state-owned as well as private com-panies involved in deforestation, mining, resource extraction, and mega-infrastructure projects, bypassing constitutional rights to prior consulta-tion, and close institutions and programmes that seek to protect the territo-rial rights of indigenous peoples.

As I have argued elsewhere (Horn 2018), a natural consequence of con-tinuities in the logic of economic development based on extraction and territorial dispossession is the continued movement of indigenous peoples towards cities or the urbanisation of previously remote, rural indigenous territories. Indeed, throughout the last ten years Bolivia and Ecuador increasingly experience the urbanisation of their rural hinterlands.[12] In Bolivia, new human settlements, housing workers engaged in the extractive industries or agricultural sector, have emerged along new transport and economic corridors crossing through the country's Amazonian and Andean regions. In the period between 2001 and 2012, 74 new urban settlements, with each being home to approximately 2,000–20,000 inhabitants, appeared on Bolivia's map (INE 2014). Similarly, in 2011 President Correa announced the Millennium Cities programme, which promoted the construction of new towns near newly emerging oilfields or other resource-extraction centres (for a critical discussion see Wilson and Bayón 2015).

Considering ongoing trends of indigenous rural-to-urban migration as well as the urbanisation the countryside, it is surprising that little research has so far been conducted into the implementation of the indigenous right to the city as laid out in the constitutions. It is precisely this topic which will be addressed in the remainder of this book.

Conclusions

The central point running through this chapter is that indigencity is not simply a static social category associated with rurality, social exclusion, and non-Western traditions that was established by the colonisers and subsequently taken up by postcolonial regimes, leading to a situation of coloniality. Indigeneity is a much more dynamic social category. Since colonialism and until the present, indigeneity describes people of different ages, genders, and socio-economic backgrounds who live both in the countryside and in urban areas, and share the fact that they self-identify as indigenous. Until now, indigeneity was mainly studied in rural areas where it shifted from a category of exclusion to a political-legal category of rights and entitlements. Indigenous peoples themselves initiated this process through contestation, resistance, and decolonisation efforts until the periods of modernisation, neo-liberalism, and (post-)neo-liberalism, when governments, international organisations, and NGOs incorporated indigenous demands into rights-based agendas. Even so, governments barely implemented these rights because of other political and economic priorities.

Even more importantly, this chapter has discussed how, why, when, and for what reasons indigenous peoples moved to the city, as well as how urbanisation arrived on their doorstep in rural territories. In a context where rural indigenous groups were granted specific rights, urban indigenous peoples remained outlawed and trapped in a situation of exclusion, marginalisation, and discrimination. Yet urban indigenous peoples still managed to preserve, revitalise, and reinvent their identity as well as to develop their own political voice. Urban indigenous peoples are, hence, not passive victims of exclusion and coloniality. They are active agents of social and political change who struggle for decolonisation and full inclusion into urban society. This was particularly evident during the insurgent uprisings in La Paz and El Alto in Bolivia and Quito in Ecuador which led to the ouster of national governments, and to the election of new governments which, as part of constitutional reforms, recognised what is here referred to as the indigenous right to the city. The indigenous right to the city acknowledges that urban indigenous peoples' interests and needs should be considered in interventions in urban space and that indigenous organising modalities should be respected in participatory processes.

Yet, until this point, little research has been conducted on what the indigenous right to the city means in practice in distinct urban contexts. Addressing this gap requires an analytical framework with which to analyse the practices of key actors involved in the implementation of the indigenous right to the city. The next chapter starts by developing such a framework.

Notes

1 It is important to distinguish the term indigenous from the term 'Indian.' The term 'Indian' was introduced by Christopher Columbus following the belief that he had reached the Indian Ocean. Even though he soon recognised that he had not arrived in the Indies but in the Americas, the name 'Indian' was subsequently used by the colonisers to name native populations and to position them at the lowest level of a newly established ethno-racial hierarchy. It is therefore a term associated with racism, discrimination, and exclusion. In contrast, indigenous is a more contemporary social category to which people self-ascribe (for a detailed discussion, see Assies 1994; Canessa 2006). Considering these different connotations, I refer to the term 'Indian' only in quotation marks and use the term indigenous when appropriate.
2 The *encomienda* system was established by Queen Isabella I of Castile in 1503 and was a system of forced labour. The Spanish crown entrusted colonisers (*encomendadores*) with rights to control certain areas. The *encomendadores* collected tributes from 'Indians' living in these areas (Platt 1982; Harris 1995). They also controlled the labour of 'Indians,' who were often forced to work, under slave-like conditions, predominantly in the sectors of agriculture or mining (Quijano 2006).
3 The concept of social exclusion originally derives from research on multidimensional deprivation and poverty in countries of the global North and South (Hickey and du Toit 2007). It describes the situation of 'individuals or groups that are wholly or partially excluded from full participation in the society within which they live' (de Haan 1999: 2). According to this definition, social exclusion has two dimensions: First, it describes a particular lived experience of multidimensional deprivation; second, social exclusion emanates from the particular practices of people. The concept of social exclusion has been increasingly used to describe the lives and historical treatment of indigenous peoples (see Wade 2010).
4 For a more detailed history on blacks in Latin America see Klein and Vinson (2006) or Wade (2010).
5 Similar to the encomienda system, the haciendas represent large landed estates in which labourers – often 'Indians' – were directly employed by the hacienda owners (mainly *criollos*). Semi-feudal working conditions prevailed (for a detailed discussion see Keith 1971).

6 Previous research on Ecuador illustrates this point. Following the 1964 Agrarian Reform Law, indigenous peoples in highland Ecuador legally owned less than three per cent of cultivable land. Similarly, the 1972 reforms did not produce significant socio-economic improvements as only 20.1 per cent of transferred land was suitable for agricultural cultivation (Korovkin 1992).

7 The *ayllu* is an ancient form of community organisation and local governance of the Quechua and Aymara peoples in the Andean region of Latin America. *Ayllus* functioned prior to the Inca period, throughout the colonial period and continue to exist today (Platt 1982).

8 Indigenous movements exercised pressure not only outside but also within governments. They established their own political parties (e.g. Pachakutik, which was affiliated with Ecuador's CONAIE, and the *Movimiento Revolucionario Túpac Katari de Liberación* – MRTKL, which was affiliated with Bolivia's CSUTCB-TK). For a discussion of the influence of indigenous parties, see van Cott (2008).

9 For a detailed discussion of rights-based approaches to development see Hickey and Mitlin (2009) and Molyneux and Lazar (2003).

10 In Bolivia alone, 30,000 miners, who were predominantly of indigenous descent, lost their jobs as the government withdrew from the country's largest public mining enterprise – the Bolivian Mining Corporation (COMIBOL) – leading to the shutdown of nearly all COMIBOL-run mines (Torrico Foronda 2011).

11 Lefebvre introduced the right to the city in response to social uprisings in Paris in 1968. Dismissive of processes of social polarisation and capitalist urban transformation occurring in cities like Paris since the late 19th century, Lefebvre (1968, 1991) argued that urban popular classes had been pushed out from central parts of the city to the urban periphery where they lacked access to services, economic opportunities, and power to influence political decisions affecting their city. In this context, Lefebvre argued that the popular classes should rise up and claim their right to the city which he defined as a collective cry and demand for renewed access to urban life. Popular movements across the world have since then integrated the right to the city into their distinct urban struggles.

12 Similar trends have also been observed in Brazil by scholars such as Roberto Monte-Mor (2018), Ana Claudia Cardoso *et al.* (2018) and Rodrigo Castriota and Joao Tonucci (2018).

3 Translating indigenous rights to the city

Introduction

So far, this book has discussed indigeneity through the lens of urbanisation, focussing on why indigenous peoples increasingly moved to cities and how urban indigeneity was defined and redefined by different people who operated in different historical and political contexts. It has also demonstrated how these different persons, whether urban indigenous residents, government officials, international donors, or church representatives, intended to preserve or change understandings of indigeneity and indigenous rights in such a way as to align them with their own interests and motivations. In other words, these people exercised agency, defined as the 'ability or capacity of an actor to act consciously and, in doing so, to attempt to realise his or her intentions' (Hay 2002: 94). These persons can, hence, be conceptualised as social actors who operate within a specific structural environment characterised by specific rules and norms but who 'are not defined by their conformity to rules and norms, but by a relation to themselves, by their capacity to constitute themselves as actors, capable of changing their environment and of reinforcing their autonomy' (Touraine 2000: 902).

This chapter now shifts the focus to urban planning and policy – the technical and political processes concerned with defining, regulating, and managing urban land use, property rights, social and economic infrastructure, and service delivery (Hall and Tewdwr-Jones 2010). In this book, urban planning and policy practice is considered as a key domain in which decisions are made on the translation of constitutional rights such as indigenous rights to the city. Like the processes which led to the ratification of Bolivia's and Ecuador's new constitutions, the translation of indigenous rights to the city into public policy and urban planning practice is conceptualised in this book as a process in which a variety of social actors are involved. A key point made in this book is that the practices of neither government authorities nor members of indigenous communities are monolithic. Rather, this book challenges dualistic representations of state-led planning practices as leading to outcomes of exclusion and dispossession of indigenous peoples, and of everyday practices of members of indigenous communities as leading to emancipatory and decolonial planning solutions. In its understanding of the urban policy and planning process, the book closely follows James Scott (1999: 256) who states:

> Any attempt to completely plan a village, a city, or for that matter a language is certain to run afoul of the same social reality. A village, a city, or a language is the jointly created, partly unintended product of many, many hands. To the degree that authorities insist in replacing this ineffably complex web of activity with formal rules and regulations, they are certain to disrupt the web in ways they cannot possibly foresee.
>
> (Scott 1999: 256)

In line with such a perspective, it is argued in this book that social actors involved in government institutions and members of urban indigenous communities are both highly diverse, hold distinct interests, and engage in different and potentially contradictory and conflicting practices which lead to policy and planning outcomes which are impossible to foresee. What analytical approaches are required to shed light on the underlying interests and motivations and associated practices of the various social actors involved in translating the indigenous right to the city? This chapter aims to answer that question by fleshing out an analytical framework for the study of urban indigeneity as a policy and planning category.

The first section of this chapter focusses on the role of social actors involved in government institutions who define how and to what extent specific constitutional rights are translated into legislation, public policy, and urban planning interventions. Translation is conceptualised as an act which is influenced and inspired not only by legal texts (such as constitutions) but also, and perhaps more importantly, by social actors' personal views, by the demands of individuals and the groups they represent, and by the specific political and institutional environments that shape their work. The chapter's second section focusses on the important role of urban indigenous residents and community-based organisations (CBOs) in influencing the urban policy and planning process. Urban indigenous residents are conceptualised in this book not as passive recipients or target groups of interventions but as social actors who seek to realise their specific interests and motivations and claim their indigenous right to the city through a portfolio of bottom-up practices. To identify and critically analyse urban indigenous resident's interests, motivations, rights-based claims, and associated practices, this chapter draws on insights from asset-accumulation frameworks and conceptual work on tactics. The third section summarises the analytical framework developed in this chapter. It also briefly returns to questions of methodology and discusses how different data collection and analysis methods were combined to identify and critically examine the practices of different social actors involved in the translation of the indigenous right to the city.

An urban governance perspective

While indigenous rights to the city are enshrined in Bolivia's and Ecuador's constitutions, the translation of these rights into urban policy and planning practice is by no means guaranteed. As Bent Flyvbjerg (2003: 325) states: 'Whereas constitution writing and institutional reform may often be essential to democratic development, the idea that such reform alters practice is a hypothesis, not an action.' The translation of constitutional content is, hence, a process whose outcomes may be good or bad in specific instances. This book explores what is done by different social actors operating within government institutions to translate the indigenous right to the city. Such an endeavour requires identifying how translation is occurring, who is involved in the act of translation, and which factors influence this process.

The policy and planning process

The translation of constitutional rights is understood in this book as a process undertaken in at least two distinct stages – policymaking and planning. Policymaking refers to the formulation of laws and public policies which further define the contents of specific rights and outline how these should be delivered to specific target groups (de Leon and de Leon 2002; Gacitúa-Marió and Norton 2009). An important question to consider at this stage is: How is the indigenous right to the city incorporated into the laws and specific sector policies?

Following the policymaking stage, the planning stage defines how specific laws and sector policies should be operationalised and implemented (Campbell and Fainstein 2003; Moser 1993). The planning stage determines the roles and functions of institutions responsible for operationalisation and implementation; the allocation of human, financial, and technical resources for relevant policy and planning interventions; and the specific programmatic guidelines that determine how the indigenous right to the city should be addressed in practice in national and local contexts.

Actors involved in urban governance

Constitution writing and the translation of constitutional contents into public policies and urban planning are separate processes which involve different institutions and social actors (de Leon and de Leon 2002). Bolivia's and Ecuador's constitutional assemblies were mainly composed of national government officials who represented different political parties (Becker 2011; Postero 2013). In the context of Bolivia, the constitutional assembly also involved members from a variety of social movements (Schavelzon 2013). In contrast to the process of writing constitutions, the translation of constitutional rights into public policy and urban planning practice is influenced not only by national governments but also by a variety of different actors involved in urban governance. Since the 1970s, the role of national governments in the planning and management of cities in Latin America and elsewhere has declined (Gilbert 2006; Kothari 2005; Mitlin and Satterthwaite 2013; Pierre 1999). In a context of globalisation, decentralisation, new public management, and neo-liberal market reforms, responsibilities around city planning increasingly involve a portfolio of institutions, including international donors, national and local governments, private enterprises, and CBOs. Local governments play a particularly important role in the urban policy and planning process. As I have discussed elsewhere, local government approaches to planning changed from focussing on urban administration in the 1970s (with emphasis on control of land planning, tax collection, and delivery of some services) to urban management in the 1980s (with a focus on the effective and efficient provision of services and citizen engagement), to urban governance since the 1990s (Horn *et al.* 2018). Urban governance requires local governments to be more accountable and transparent; more adroit at contributing to the transformation of the diverse forms that local democracy assumes; and more capable of involving proactively different public, private, and civil society actors who operate at different levels, such as at the community or neighbourhood, city, national, regional, or global level.

Urban governance processes, and the institutions and actors involved in these processes, vary depending on context. In Bolivia and Ecuador national governments have promoted reforms which resemble a 'return of the state' (Elwood *et al.* 2016). This means that core programmes, interventions, and economic activities which were previously outsourced to non-state organisations in the private, donor, or civil society sector have been reintegrated within the state (Horn and Grugel 2018). In contemporary Bolivia and Ecuador, then, national and local governments play a key role in urban governance, and it is for this reason that this book puts emphasis on the practices of authorities, planners, and policymakers who operate within the realm of national and local governments.

What influences the practices of actors involved in urban governance?

Constitutions certainly provide policymakers and urban planners with normative guidelines on how to define specific laws, public policies, and urban planning interventions. Yet, to reiterate the argument made at the beginning of this section, the translation of constitutional rights is not guaranteed as 'constitutions are often written on water' (Putnam 1993: 17). A critical analysis of acts of translation must therefore pay careful attention to gaps between constitutional rhetoric and practice. This requires moving beyond an analysis of legal texts and recognising that government officials involved in acts of translation have agency, form part of specific actor coalitions, and operate within a fragmented political and institutional environment characterised by uneven power relations.

Let us unpack each of these points in detail. The chapter started by defining people involved in the translation of the indigenous right to the city as heterogeneous social actors who hold different personal interests and motivations that influence their practices. While some government authorities may certainly define their practices in relation to rights established within current constitutions and legal documents, it is important to emphasise that others may still define their practices in relation to previously established norms. Sociologist Pierre Bourdieu (1977) refers to the influence of history as a determining factor of the practices of social actors as habitus. The habitus represents the guideline for human behaviour and action. Human beings are considered to act upon internalised 'objective' structures, norms, and values which are themselves outcomes of past events. The habitus, therefore, is 'a product of history, produces individual and collective practices – more history – in accordance with the schemes generated by history' (Bourdieu 1977: 54). Government officials involved in the translation of indigenous rights are often described as being guided by a colonial habitus. For example, despite the introduction of anti-discrimination laws and the recognition of international legislation for indigenous peoples in the neo-liberal period (see Chapter 2), government officials in different Latin American countries continued to treat indigenous peoples, independent of their rural or urban residence, like second-class residents who were considered inferior to the countries' 'white' *criollo* or *mestizo* populations, thereby reproducing patterns of ethno-racial stratification introduced by the colonisers (Engerman and Sokoloff 2000; Lucero 2004).

Other scholars draw on Bourdieu's habitus to offer a critical analysis of planning's past and present role in addressing indigenous peoples. Planning plays a significant role in implementing systems of ethno-spatial segregation and indigenous dispossession through specific ways of regulating urban space. Writing on Latin America, Jorge Hardoy (1973, 1989) reveals how the Spanish colonisers planned their towns and cities in such a way that they reproduced European political and social institutions such as the church and the town hall which were situated at central squares. 'Indians' were denied from inhabiting these spaces but forced to reside in unplanned peripheral settlements or in the countryside. Hardoy (1989) as well as other Latin American scholars such as Richard Morse (1978) and Anibal Quijano (1975) have argued that colonial planning practices were subsequently reproduced by planners working in (post)colonial governments. Guided by a colonial habitus, planners continued producing ethno-spatially divided urban spaces, with indigenous peoples often confined to living in informal settlements, squatter camps, or social housing complexes characterised by precarious housing conditions and an absence of basic public services.

Similar trends have been noted for (post)colonial and settler colonial town and city planning in other former colonies in the global North and South. Writing on North America, Australia, and New Zealand, Libby Porter (2010) offers a detailed account of how the British colonial regime used planning tools such as cadastral surveys, mapping, and zoning to establish new racially seg-regated towns and to acquire land from native population groups. She argues that the writings of John Locke, who considered property as arising from 'the labour applied to objects and land,' particularly influenced British colonial planning practices of land acquisition (Porter 2010: 56). The British consid-ered their colonies as empty or wastelands (even though they were often home to native population groups) and, therefore, as available for the application of a specific type of labour – modern, rational, and industrial rather than tra-ditional, pastoral, and indigenous labour. Once such labour was undertaken on a plot of land, it became private property and could no longer be taken away. Planning practices introduced by the colonisers were subsequently re-produced by (post)colonial regimes, reinforcing what Porter refers to as co-lonial cultures of planning. Writing on East Africa, Garth Myers (2003) also outlines how the British colonisers used specific planning strategies to create ethno-racially divided spaces which were subsequently reproduced by plan-ners working within (post)colonial governments. Such strategies included enframing (the replacing and destruction of 'unordered' precolonial urban structures), creating insiders and outsiders through racialised zoning (that is, planned central spaces for 'whites,' Asian buffer zones, and unplanned peripheries for the indigenous black population), and objectifying and con-trolling urban space (that is, through the establishment of points of obser-vation). The habitus, or better the ongoing influence of colonial and (post) colonial norms and planning practices, can influence present attitudes and practices of government officials, and must therefore be considered as impor-tant for an analysis on how planners and policymakers translate indigenous rights to the city.

The practices of government officials, policymakers, and planners are also influenced by power relations. According to Bourdieu (1986), power is exercised by certain people over others and determined by the amount of economic (e.g. money), social (e.g. connections, position in an institution), or cultural resources (e.g. university degrees) a person possesses. The more resources people have, the more powerful they are. Within government in-stitutions this means that actors in more senior level positions are often considered better able to influence, shape, and constrain the work of other actors. Such tendencies have been noted in previous studies of government institutions responsible for addressing indigenous rights. These include, for example, CODENPE and DINEIB in Ecuador or MAIPO in Bolivia. While these institutions were composed of staff who actively promoted a pro-indigenous political agenda, their practices were constrained by the fact that higher-ranking authorities in other sector ministries allocated them insufficient financial resources (Andolina *et al.* 2009; Hornberger 2000; Yashar 2005).

It is not only senior government authorities who have the power to address or withhold specific rights, resources, and services. Rather, power sits every-where and can be used and (ab)used by a variety of actors with different levels of influence. As Michael Lipsky (1980) pointed out, it is often lower tier administrative staff who play the key role in determining whether and how specific norms are addressed on the ground. Such tendencies have been reported by Gonzalo Colque (2009), who, writing on Bolivia, argued that low-tier municipal staff refused to comply with national legislation which required them to address indigenous peoples in their native languages. The power of 'street level' bureaucrats, which helps explain why some groups of society continue to experience structural discrimination whilst others benefit, is undoubtedly key for an analysis on the translation of the indigenous right to the city.

The practices of government officials, urban planners, or policymakers are also likely to be defined in relation to the actor coalitions to which they belong or which they represent. For example, government officials generally belong to different political parties that represent different interest or pressure groups which can include their electoral support base, as well as international donor organisations and private sector pressure groups. Writing on Bolivia's current government, led by President Evo Morales and the MAS party, Nancy Postero (2017) shows that the MAS has to address the interests of a very heterogenous electoral base, comprised of the poor, indigenous, and popular sectors. This requires balancing tensions between very different political agendas based on indigenous cosmovision, leftist, and populist principles. Recent research on policy and planning practice in rural Bolivia and Ecuador confirms this trend. In rural settings, government authorities are considered to focus more on the expansion of economic activities such as natural-resource extraction with the aim of increasing financial resources available for government spending on wealth redistribution and poverty-reduction. Government attempts to intensify natural-resource extraction, however, often take place on the territories belonging to indigenous communities, with authorities bypassing internationally agreed mechanisms on community rights to prior consultation (Anthias 2017; Filho and Goncalvez 2010; McNeish 2013; Walsh 2011). Addressing the interests of the poor, hence, comes at the cost of violating rural indigenous peoples' territorial and political rights.

A brief reflection on public policy and urban planning practices in cities situated elsewhere supports the argument that social actors involved in urban governance undertake interventions in ways that conform with their own political agenda as well as the interests and demands of the coalitions and communities they belong to, represent, or rely on. Writing on the 'dark side of planning,' Oren Yiftachel (2006) shows how planning authorities use land regulation as a strategic tool to selectively include or exclude specific individuals and groups from inhabiting urban space. This is illustrated through the practices of planning authorities in Israel/Palestine who declared the nomadic and collective land-use patterns of indigenous Bedouin Arabs as illegal. The government used this planning strategy to displace Bedouin Arabs from their ancestral territories and to provide Jewish settlers, who intended to occupy and urbanise this land, with individual tenure rights. Writing on Mumbai in India, Ananya Roy (2009) demonstrates how planners 'invent' states of exception in order to legalise urban land acquisition by wealthy elites or real-estate enterprises while simultaneously criminalising land acquisition by the urban poor by means of evictions. Urban policy and planning practice, hence, is not a neutral process but can serve to reinforce patterns of exclusion and dispossession of one group – the indigenous, the poor, the powerless, the underrepresented – while favouring others – the rich, the privileged, the powerful.

In brief, then, how the indigenous right to the city is translated into practice is not only influenced by legal texts such as constitutions but also depends on the personal or material interests of those in office and relates to the coalitions or communities that government authorities, planners, and policymakers belong to, represent, or rely on as well as on questions of power relations within the broader institutional environment of urban governance.

The role of urban indigenous residents

In an environment in which social actors operating in government institutions do not automatically translate constitutional content, urban indigenous residents are likely to deploy their own solutions and claim the indigenous right to the city from below. In this book, then, the indigenous right to the city should not be conflated with being only a legal, policy or planning category. Instead, it also represents an entry point to investigate how ordinary indigenous residents address their own interests and demands. As I have argued elsewhere with emphasis on low-income groups (Horn *et al.* 2018), at the neighbourhood level such bottom-up practices are likely to include involvement in CBOs and engagement in community self-help practices. At levels beyond the neighbourhood, representatives or leaders of CBOs and other organisations may, among other activities, interact with each other to further their objectives; engage in political negotiations with actors in the public, donor, or private sector; or trigger insurgent uprisings and protest. Based upon such a perspective, urban indigenous residents are considered in this book as social actors who can plan their own lives according to their interests and motivations, and who have the capacity to claim their right to the city and reshape and influence urban governance from below. It is therefore important to identify the different interests, demands, and rights-based claims of indigenous residents; how urban indigenous peoples get what they want; and which factors influence these bottom-up practices.

Chapter 2 outlined how previous research described the interests and demands of Bolivia's and Ecuador's urban indigenous peoples. For example, David Guss (2006) shows how indigenous residents in La Paz seek to revitalise their festivals and ancestral traditions within their new urban communities. Xavier Albo *et al.* (1981) and Eduardo Kingman (2012) point out that indigenous peoples in Bolivia and Ecuador mainly move to the city in search of better education and economic opportunities. Daniel Goldstein (2004) and Sian Lazar (2008) demonstrate how urban indigenous residents in the cities of Cochabamba and El Alto want, like most other urban residents, to live in secure neighbourhoods and to have access to modern urban amenities such as housing, water, roads, or electricity.

While offering important insights, such accounts provide only a partial understanding of some of the cultural, economic, and social interests, and motivations and associated rights-based claims articulated by urban indigenous residents. An analytical framework which allows for a more integrated analysis is therefore required.

Identifying what urban indigenous residents want: an asset-accumulation framework

To identify what urban indigenous residents want, this book draws on the theoretical and policy-focussed international development literature on asset accumulation (Appadurai 2004; Bebbington 1999; Carter and Barrett 2006; Moser 1998, 2009; Sherraden 1991). Approaches focussing on asset accumulation take inspiration from the work of Amartya Sen (1981) on entitlements, assets, and capabilities and Robert Chambers (1994) on risk and vulnerability. As indicated by its name, an asset-accumulation framework is mainly concerned with assets and associated strategies for accumulating assets. According to Caroline Moser (2009: 18), assets can be defined as a 'stock of financial, human, natural or social resources that can be acquired, developed, improved and transformed across generations.' In addition, more intangible assets, such as political, cultural, or aspirational capital, may further affect what people want in life (see Table 3.1 for a definition of different capital assets). Writing specifically on cultural and aspirational capital, Arjun Appadurai (2004: 10) argues that 'everyone, including the poor, express their aspirations, hopes and choices.' People's specific aspirations, interests, and demands are likely to differ between individuals and groups who profess different cultures, norms, or identities.

An asset-accumulation framework has been used as a diagnostic and analytical tool to identify what different assets people want to access. It has been applied to the study of the interests and motivations of the urban poor and also of international migrants. In a longitudinal study in a poor urban community in Guayaquil, Moser (2009) discusses the different interests and demands which poor urban residents from one neighbourhood articulated over a period of 30 years – from 1978 until 2008. During initial stages of neighbourhood consolidation, members of this community were mainly interested in accessing housing so that they could live together with their families. They often achieved this through self-help efforts. Having a house was a precondition to access other assets such as electricity, water, or sanitation – a trend which is also confirmed in other studies on poverty-reduction in cities of the global south (Satterthwaite 2008; Turner 1976). Moser's (2009) Guayaquil study further shows that the interests and demands

Table 3.1 Definition of capital assets

Asset	Definition
Physical	The stock of plant, equipment, infrastructure, and other productive resources owned by individuals, groups, businesses, or the state.
Financial	The financial resources available to people such as credits, savings, monthly wages, etc.
Human	Investment in education, health, and the nutrition of individuals. Labour is linked to investments in human capital, health status determines people's capacity to work, and skills and education determine the returns from their labour.
Social	The rules, norms, obligations, reciprocity, and trust embedded in social relations, social structures, and societies' institutional arrangements. It is embedded at the micro-institutional level (communities and households) as well as in rules and regulations governing formalised institutions in the marketplace, political system, and civil society.
Natural	The stock of environmental resources provided such as soil, water, minerals, or land. In rural communities, land is a critical productive asset for the poor; in urban areas, land for shelter is also a critical productive asset.
Political	Knowledge of existing rights, rules, and procedures that characterise a given political system. Contacts to relevant actors in urban governance.
Cultural	Dress, appearance, or specific type of education. It can also refer to people's specific habits, such as what food, music, and behaviours they identify as appropriate or otherwise.

Source: Elaborated by the author, drawing on Appadurai (2004) and Moser (2009).

of members of this poor urban community changed over time, shifting from housing to accessing community infrastructure such as roads. Demands for assets also differed between members of this community – men expressed different demands for work than women; children had other aspirations for education than their parents.

In addition to gender-based or intergenerational differences, people's asset-based demands vary depending on the location of their residence as well as upon the specific political context they confront. A study of the asset-based demands of group of migrants from Ecuador living in Barcelona by Caroline Moser and myself (2015) found that migrants value citizenship or temporary residency rights as particularly important because access to such political assets is associated with opportunities to access other assets, such as housing, bank accounts, or education. Similarly, Jorge Ginieniewicz (2015) emphasises the importance of political capital for Argentinean migrants living in Spain. Only through developing an understanding of the existing rights, rules, and procedures of their host society could this group of migrants access other important assets, such as a job or residency permit. Also writing on migration, Cathy McIlwaine (2011) shows how Latin American migrants in London express the need to adjust their human capital, such as English-language skills, to access crucial assets such as housing or education.

Overall, then, an asset-accumulation framework demonstrates that people, whether from different communities or from the same community and of similar background, are likely to articulate different aspirations and demands for assets which can also change over time, generation, and space. Further, demands for one asset – such as education or citizenship status – are often related to aspirations for other assets – such as housing or better work. An asset-accumulation framework is therefore useful for the analysis of what different urban indigenous residents want. As will be demonstrated in Chapter 5, an asset-accumulation framework helps to break away from a discourse on indigenous communities as being relatively homogenous. Instead, it can shed light on how indigenous communities are heterogeneous and, at times, internally divided with different members (of different age, gender, and levels of power) expressing distinct and, at times, conflicting and contradictory interests, motivations, and rights-based claims.

Contesting and negotiating the indigenous right to the city from below

An asset-accumulation framework is useful not only for the identification of what different people want in life. It also helps in generating an understanding of how people get what they want and, as such, helps capturing how people claim their indigenous right to the city. A central assumption of asset-based frameworks is that assets provide people not only with access to a set of resources but with the capability to be and act (Bebbington 1999). Amartya Sen (1997) emphasises that the acquisition of assets is associated with empowerment as it enables people to confront and negotiate with authorities and to influence or change the political environment that governs them. Opportunities to access and accumulate a portfolio of assets depend at least on two factors – agential and structural. In terms of agential factors, the more assets a person already possesses, the more likely this person is to be able to contest and negotiate access to other assets (Moser 2009). Those community members who obtain the relevant social and political capital (such as connections with social actors in urban governance and knowledge of the political system that surrounds them) are most likely to obtain leadership positions and to play a central role in contesting and negotiating access to a portfolio of assets for themselves and, at times, for the communities they supposedly represent. Community leaders, like actors in urban governance, can also abuse their powerful position and further exacerbate problems such as exclusion. For example, Nicola Banks (2014) demonstrates in a study of urban poor communities in Dhaka, Bangladesh, how local leaders – disproportionately richer than other residents and well connected to local political party leaders – use their position for purposes of self-enrichment. Banks also illustrates how community leaders, who hold an interest in preserving their powerful position within the community, selectively reward supporters while denying other residents' access to crucial services and resources. Such interpretations of communities and community-based bottom-up practices are key to analysing the indigenous right to the city. They help in de-essentialising understandings of indigenous communities who are relatively homogenous, live in harmony with each other, and whose collective practices are often considered to lead to emancipatory social change (Jojola 2008; Zibechi 2010). While bottom-up practices by members of organised indigenous communities may certainly lead to social change, it is also important to recognise that indigenous communities are ultimately characterised by uneven power relations where exclusion and dispossession can be reinforced by its own members.

The capacity of a person or community to access assets also depends on a variety of structural and institutional factors (Moser 2009). Structural factors may include existing norms, regulations, or planning processes which may enable or constrain people's opportunities to access their aspired portfolio of assets. Furthermore, whether people can access specific assets also depends on the willingness of different actors operating in institutions associated with urban governance. Scholars deploying an asset-accumulation framework, therefore, started developing a set of theoretical, methodological, and practical guidelines on what could be done to generate a political environment which is responsive to the asset-based demands of specific individuals and groups. For example, writing on urban poverty-reduction in the context of the Millennium Development Goals, Alfredo Stein and myself (2012: 669) argue that policymakers should rely on three generations of asset-policies. First-generation policy strategies should provide the urban poor with access to basic services such as housing, water, or sanitation. Second-generation strategies should further ensure the consolidation of these assets and prevent their erosion (through the provision of citizen rights or institutional accountability). Third-generation strategies should maximise the links between different interdependent assets.

Scholars working on urban indigeneity in settler colonies of the global North have also started to develop a set of ideal-type models of what could be done by actors in urban governance – a realm generally associated with reproducing (post)colonial patterns of exclusion and indigenous dispossession (Hardoy 1989; Myers 2003; Porter 2010; Yiftachel 2006) – to better address indigenous interests, demands, and rights-based claims. Leonie Sandercock (2003), for example, proposes a model of cosmopolitan cities which emphasises the importance of addressing the diverse interests of different urban residents, including indigenous peoples. Inspired by philosophical work on multiculturalism (Kymlicka 1995) and the communicative turn in planning (Healey 1997), Sandercock emphasises that policymakers and urban planners should – instead of being implementers of rights – act as facilitators who engage urban residents of different age, ethnicity, gender, or socio-economic status in decision-making processes. In their work on planning for coexistence, Libby Porter and Janice Barry (2016) also emphasise the need for better collaboration between planners operating in urban governance and representatives of indigenous communities. They argue that interactions between government authorities and indigenous communities take place in contact zones, spaces in which collaborative efforts around the translation of indigenous rights and the recognition of indigenous governance occur. But collaboration is often limited within contact zones that are characterised by uneven power relations, with indigenous peoples often only representing visitors invited to a world of urban governance which remains unchanged and taking part in a skewed co-governance process which is often characterised by 'depoliticization, idealized consensus building and an inclusionary politics stripped of its attention to deep and persistent colonial relations of power and dispossession' (Porter and Barry 2016: 189). Overcoming these structurally enshrined problems, then, requires a change in attitude among planning professionals. Ultimately, planning professionals should develop intercultural capacity and embrace conflict, paying particular attention to the teachings and planning practices of indigenous peoples themselves. Others consider indigenous planning as a practical approach which focusses on indigenous self-determination and autonomy, and which emphasises that members of indigenous communities should reclaim their historic, contemporary, and future-oriented planning approaches (see also Jojola 2008; Prusak *et al.* 2016).

While such approaches offer important insights on how to potentially address the indigenous right to the city, this book moves beyond a focus on *what could be done* to address these historically marginalised groups. Instead, it puts emphasis on exploring what indigenous residents are *actually doing* to fulfil their specific interests and rights-based claims within the political environment that governs. In doing so, it conceptualises the practices of indigenous residents as tactical manoeuvres which play with, or potentially seek to alter, existing institutional structures and rules. In its definition of tactics, the book follows French critical theorist Michel de Certeau (1984: 37):

> The space of tactic is the space of the other. Thus it must play on and with a terrain imposed on it and organised by the law of a foreign power. It is a manoeuvre 'within the enemy's field of vision'. It operates in isolated actions, blow by blow. It takes advantage of 'opportunities' and depends on them.

Depending on the structural and political environments, indigenous residents are considered to rely on a portfolio of different tactics. Faranak Miraftab (2009) offers a useful classification of tactics as taking place either in invented spaces (established by civil society groups to directly challenge the status quo) or invited spaces (platforms for participation and citizen involvement designed and controlled by states). Tactics in invented spaces are not formally recognised by actors in urban governance and often take place in a context of denial or exclusion from rights or services. Such tactics can refer to self-help action. For example, ethnographic accounts on La Paz and Quito outline how urban indigenous peoples, in a context in which municipal governments did not provide them with services or support in the neoliberal period, relied on self-help efforts such as the *minga/mita* (collective work schemes) to construct houses for indigenous community members who arrived at the city from the countryside (Albo 2006; Kingman 2012). In addition to self-help practices, Miraftab (2009: 44) also discusses other tactics which take place in invented spaces; she defines those as insurgent planning:

> A range of actors may participate in insurgent planning practices: community activists, mothers, professional planners, school teachers, city councillors, the unemployed, retired residents, etc. Whoever the actors, what they do is identifiable as insurgent planning if it is purposeful actions that aim to disrupt domineering relationships of oppressors to the oppressed, and to destabilise such a status quo through consciousness of the past and imagination of an alternative future.

As outlined already in Chapter 2, research on Quito's, La Paz's, or El Alto's indigenous peoples discussed how urban indigenous insurgency led to the removal of Bolivia's and Ecuador's governments (Becker 2011; Lazar 2008; Revilla 2011; Zibechi 2010). Unlike in invented spaces, tactics within invited spaces are formally recognised and normally comply with rules, norms, and procedures established by actors operating within the realm of urban governance. Tactics in invited spaces can refer to, among others, the involvement of indigenous residents in registered demonstrations, processes of political negotiation, or engagement in participatory processes. As noted by Steven Robins *et al.* (2008), the rules, norms, and procedures established by actors and governments differ in local contexts. Particularly in Latin American cities official rules, norms, and procedures rarely match models of Western democracy but are often characterised by patron-client relations – i.e. the exchange of votes and political support for favours and services between urban residents and authorities (Auyero 2000; Moser 2009). Patron-client relationships have been reported in Sian Lazar's (2004, 2008) work on the city of El Alto in Bolivia where, during pre-election periods, urban indigenous residents often supported one or multiple party candidates in order to receive favours (i.e. money or infrastructure projects for their communities). Considering that urban indigenous peoples, at least according to Bolivia's and Ecuador's new constitutions, should be involved within participatory and decision-making processes around the indigenous right to the city, it is important to identify and critically analyse to what extent and how urban indigenous peoples deploy different tactics when engaging not only in invented but also in invited spaces.

In brief, then, an analytical framework which takes inspiration from conceptual work on asset accumulation and tactics not only provides important ideas on how to study the interests, motivations, and rights-based claims of indigenous residents. It also provides useful entry points to identify and critically analyse different tactics taking place in invited and invented spaces, which urban indigenous peoples can potentially use to claim their indigenous right to the city.

Conclusions and reflections on methods

This chapter has developed an analytical framework which conceptualises the translation of the indigenous right to the city into public policies and urban planning as a complex process shaped by a variety of social actors. It started by reflecting on the role of government authorities, planners, and policymakers who operate in the realm of urban governance and are responsible for translating constitutional content into laws, policies, or urban planning interventions. It highlighted the fact that constitutional contents on the indigenous right to the city are likely to be only one of many factors which influence the work of people involved in this process. In addition, social actors involved in urban governance may define their translation practices in such a way that they align with the interests of individuals and groups whom they represent or with their own personal views. Translation practices may also be influenced by the wider political, institutional, and structural environment that shapes their work as well as power relations.

It was also suggested that public policies and urban planning interventions can best be studied in relation to the impact they have on relevant target groups – urban indigenous residents. Like people involved in urban governance, urban indigenous peoples are social actors whose action is guided by their specific interests and motivations. An asset-accumulation framework and conceptual work on tactics are particularly helpful in generating an understanding of what urban indigenous peoples want and how they claim their indigenous right to the city within the specific political environment in which they operate.

The analytical framework fleshed out in this chapter informed the choice of methods that were used to gather empirical material on the translation of indigenous rights to the city in La Paz and Quito. Chapter 1 already introduced the qualitative, comparative research methodology that was used for this study and offered some background on methods. To identify and critically analyse *which* practices different social actors rely on to translate indigenous rights to the city into public policies and urban planning practices and *how* and *why* these practices occur in the first place, it was decided to combine different methods of data collection and analysis. For example, observations of the political negotiation approach of indigenous leaders combined with semi-structured interview and focus-group material which captures understandings of indigeneity and the associated interests, demands, and rights-based claims of leaders but also of the members of their CBOs help to explore *how* and *why* indigenous leaders realise specific practices and tactics. Methods were also deployed in such a way that they can capture the impact of power relations and structural factors (e.g. in a historical, institutional, or legal context) on the practices of different actors. For example, the analysis of documents such as constitutions and laws which define indigenous rights, combined with material from interviews and observations of the practices and interactions of different government officials, helped to develop an understanding of *who* decides *how* specific indigenous rights are understood and addressed in practice. Before offering a detailed comparative analysis on the translation of the indigenous right to the city, the next chapter sets out some important contextual background on the cities of La Paz and Quito.

4 The making of two indigenous cities

Introduction

With most of the indigenous peoples discussed in this book either affected by urbanisation or directly living in urban areas, cities are not only important sites in which indigenous political struggles unfold but also key places in which specific indigenous rights-based claims must be addressed by decision makers involved in policy and planning practice. Focussing on La Paz and Quito, this chapter offers some contextual background on the two cities which represent the case studies of this book. The chapter provides a history of indigenous urbanisation and discusses the ethno-spatial make-up of both cities, highlighting how patterns of ethno-racial and socio-economic segregation established during the colonial period persist to a certain degree until the present context. The chapter also provides an overview of the various national and local government authorities involved in urban policy and planning practice in these cities.

Chukiyapu Marka/La Paz: two names for one city

La Paz is one of Latin America's most spectacular cities surrounded by some of the highest peaks of the Andes (see Figure 4.1). It is located at a breathtaking altitude of over 3,500 metres above sea level. To reach La Paz, a visitor must first navigate through the busy streets of El Alto, another indigenous megacity situated on the Bolivian Altiplano plateau and home to approximately 900,000 inhabitants who are predominantly of Aymara background (Gill 2000; Lazar 2008). Arriving at a steep valley edge in the central Ceja neighbourhood of El Alto, one gains a first bird's-eye view of the concrete jungle that is La Paz. From there, one travels downwards through the city's hillside neighbourhoods, called *olladas*, which are home to a majority indigenous population. Once at the bottom of the north side of the valley, the visitor arrives in the city's colonial centre and continues the journey through the business district of El Prado, later passing through bohemian Sopocachi and San Jorge, and subsequently descending further down into the valley to arrive in the affluent

Figure 4.1 View of southern La Paz and the Illimani mountain.

and suburban Zona Sur. These areas of La Paz were traditionally home to a largely prosperous *criollo* (of Spanish descent) and *mestizo* (of mixed descent) population. This trend is slowly changing as, during the last decade, a small number of more affluent indigenous residents, mainly belonging to the newly emerging Aymara business elite, started buying real estate in these more 'white' parts of town (Maclean 2018). After passing through the Zona Sur, the visitor ascends to the steeply located, peripheral hillside districts of Chasquipampa and Ovejuyo, leaving behind the more affluent part of the city and entering neighbourhoods that are home once again to a mainly indigenous population.

La Paz, then, is a city with a clearly racialised urban geography. The ethno-racial, socio-economic, and demographic characteristics of this city can best be understood by reflecting on its history. Before the Spanish conquest in the early 16th century, the area of today's La Paz was an Inca settlement known as Chukiyapu Marka. Indigenous peoples have not forgotten their historical ties to this city and continue to call it by its precolonial name. In 1573, the Spanish colonisers destroyed Chukiyapu Marka and built their own city on its ruins. This city was called Our Lady of Peace (*Nuestra Señora de La Paz*) or, in short, La Paz. While there is no official record of the number of indigenous people residing in the new colonial city during the time of the Spanish conquest, Guss (2006) reports that in 1573 around 5,820 indigenous people and only 260 Spaniards lived in the territory of today's La Paz. The proportion of indigenous people living within the city has varied throughout the centuries but, to this day, La Paz still retains a strong urban indigenous presence. According to the most recent national census data, in 2012 around 220,000 of La Paz's 760,000 residents identified themselves as belonging to an indigenous nation or people (INE 2014). Eighty-nine per cent of those who identify themselves as indigenous now claim to be of Aymara origin, with the remainder being mainly of Quechua and Guarani origins.

Drawing on colonial city-planning practices common for the Latin American region (see Chapter 3), the colonisers excluded indigenous peoples from residing in the inner city – a privilege which was reserved to the Spanish and, in later periods, *criollo* and *mestizo* elites. Guided by 15th-century European models of architecture and planning, the colonisers structured the inner city according to a grid model which contained at its centre the main square – initially known as *Plaza Mayor* (known today as *Plaza Murillo*) – where the town hall and cathedral of La Paz are located. Until the present, this part of the city, referred to as *Centro* or *Centro Colonial*, represents the political heart of Bolivia and it is here that most national government ministries, the presidential palace, and municipal government offices can be found.

With the arrival of the colonisers, indigenous peoples were employed to build the new colonial city. They also worked as servants in 'white' middle- and upper-class households. Indigenous peoples were forbidden to inhabit or own property within the colonial city. Instead, they lived in *barrios indios* – neighbourhoods for the native population – or in rural areas surrounding the city. The first *barrios indios* were situated on the steep hills of the north-western parts of the La Paz valley; today, these form part of the more indigenous neighbourhoods of San Pedro, Villa Victoria, and Max Paredes. Early colonial city maps depict *barrios indios* as small clusters of houses which lacked basic passageways or roads.

For the first 350 years of its existence, La Paz served as a transport hub and administrative centre for the coordination of agricultural and tin-mining activities in the nearby countryside. In 1830, the city became home to the Higher University of San Andres (*La Universidad Mayor de San Andrés*, UMSA), one of Bolivia's largest and most respected universities. In 1889, La Paz became the administrative seat of Bolivia's national government. However, until the present, La Paz did not have official status as the capital city. Instead, Sucre, which contains the seat of Bolivia's constitutional court, remains Bolivia's capital. In the late 19th and early 20th centuries, La Paz also became Bolivia's national centre for industrial textile production.

Despite being Bolivia's educational, economic, and political centre, La Paz remained for some time a relatively small Andean city. In 1900, for example, it had only 50,000 inhabitants (Guss 2006). This changed rapidly from the mid-20th century onwards. In 1960, La Paz was already home to approximately 360,000 inhabitants. By 1975, the population had nearly doubled to 600,000, while by 2012 the city had around 760,000 inhabitants.

A large proportion of new residents are of indigenous descent. Large-scale indigenous migration into the city began after the 1953 Agrarian Reform Act, which abandoned the *hacienda* (large landed estate) system and allowed indigenous peoples to move freely throughout the country (Dunkerley 2007). Indigenous peoples mainly came to La Paz in search of a decent living. Push factors included failed land reforms in the countryside and national economic crises which made it difficult for indigenous peoples to sustain a peasant lifestyle (Albo *et al.* 1981; Lazar 2008). Pull factors included educational and economic opportunities available in the city.

The mass migration of indigenous peoples to La Paz led not only to an increase in the city's population but also to unplanned and rapid urban expansion. Between the 1960s and 1990s, La Paz more than doubled in physical and population size (Arbona and Kohl 2004). Figure 4.2 illustrates the physical growth of the city between 1976 and 1996. Urban growth and expansion followed previously established ethno-spatial segregation patterns. La Paz's more 'white' and affluent *criollo* and *mestizo* elites increasingly moved away from the colonial city centre to lower parts of the valley, with new middle- and upper-class neighbourhoods emerging, first in Sopocachi, San Jorge, and Miraflores, and later in the Zona Sur, especially in the neighbourhoods of Obrajes, Calacoto, and Achumani. These closely resemble middle- and upper-class neighbourhoods that can be found throughout Latin America and elsewhere in the world. Indigenous migrants hardly located to these parts of the city but, rather, occupied land and self-constructed makeshift houses in the growing unplanned urban peripheries situated along the steep terrain on the north-western or north-eastern sides of the valley of La Paz, close to the first *barrios indios*. Settlement through squatting, occupancy, and incremental self-help housing was, of course, not unique to La Paz's growing urban indigenous population but has been observed for low-income groups across Latin America (Turner 1968, 1978). Indigenous migrants who arrived from the same rural areas normally settled in the same street or neighbourhood, thereby reproducing their communities within the city (Albo *et al.* 1981). By the 1970s, La Paz's northern peripheries were already densely populated and newer generations of indigenous migrants started to settle in La Paz's south-eastern periphery, especially in the neighbourhoods of Chasquipampa and Ovejuyo, or on the plateau west of La Paz which represents today's municipality of El Alto (Arbona and Kohl 2004).

Urbanisation patterns in La Paz not only replicated previously established ethno-spatial segregation patterns. Socio-economic inequalities between more indigenous and non-indigenous parts of the city also persist. For example, data provided by the municipal government of La Paz (2006) reveal that the indigenous population rarely resides within the urban core but lives mainly in peripheral urban neighbourhoods situated on the hillside slopes of the La Paz valley (see Figure 4.3). Compared to more central, non-indigenous parts of the city, these neighbourhoods lack crucial public

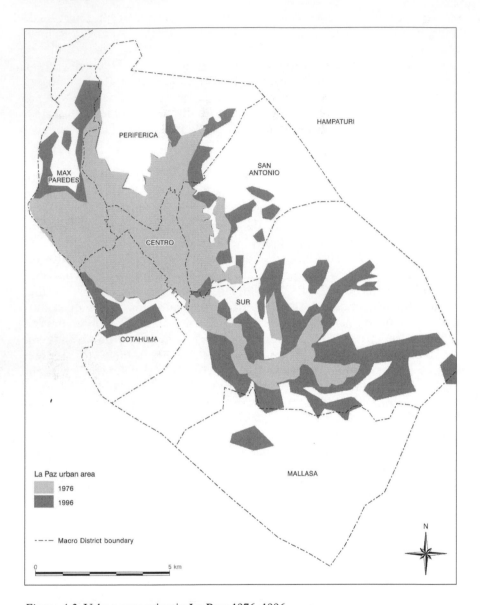

Figure 4.2 Urban expansion in La Paz: 1976–1996.

amenities and services such as access to sanitation (see Figure 4.4). Houses within these more indigenous neighbourhoods are also likely to be constructed on earth floors. In addition, there remain crucial differences in the levels of education between indigenous and non-indigenous neighbourhoods. For example, illiteracy rates remain significantly higher in La Paz's indigenous peripheries than in lower-located and more central neighbourhoods with fewer indigenous residents (see Figure 4.5).

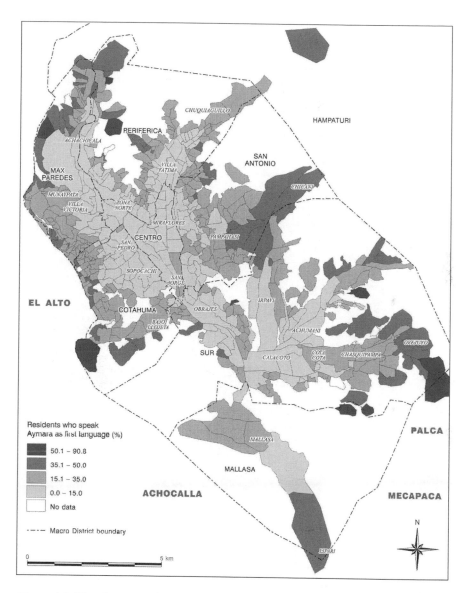

Figure 4.3 First-language Aymara speakers in La Paz.

It is of course important to recognise that data from the 2001 municipal census cannot provide a completely accurate description of current trends relating to ethno-spatial and socio-economic inequalities within the city. In fact, between 2001 and the present, local and national government authorities undertook a variety of interventions, including the introduction of new housing and public-service schemes within more indigenous neighbourhoods (see also Chapter 5). At the same time, though, an emerging trend of social

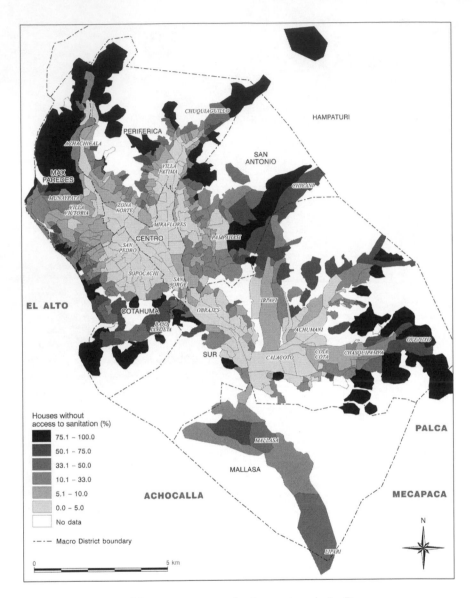

Figure 4.4 Houses without access to sanitation systems in La Paz.

upward mobility among a small group of Aymara business elites that are involved in the city's commercial sector can be noted (Maclean 2018; Tassi *et al.* 2013). These shifts notwithstanding, findings from the 2001 municipal census seem illustrative of the current situation. For example, interviews conducted for this book with government representatives, indigenous residents, and community activists in 2012 and 2016 found that patterns of ethno-spatial,

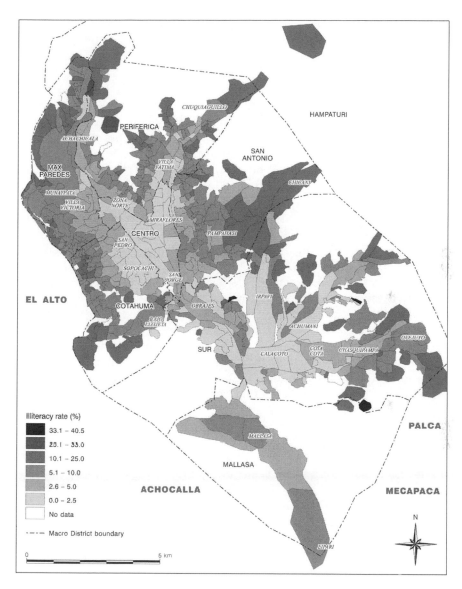

Figure 4.5 Illiteracy levels in different neighbourhoods in La Paz.

and related socio-economic, divides persist between the more 'white' parts of La Paz and the more 'indigenous' Chukiyapu Marka. These divisions can best be understood through Anibal Quijano's (2000, 2007) concept of coloniality of power. They represent a continuation of the dynamics of social exclusion and of spatial as well as socio-economic and political inequality initially established by the colonisers. Maria, an indigenous migrant who moved from a small village near Lake Titicaca to the neighbourhood of Ovejuyo, offered a succinct description of this phenomenon (Interview, 11 November 2016): 'Like our Aymara ancestors who lived in Chukiyapu Marka under the colonial regime, being Aymara at present means living on the periphery, being a second-class citizen, and receiving second-class services.'

In addition to indigenous urbanisation induced by rural-to-urban migration, La Paz also experienced natural indigenous urban growth and physically absorbed previously rural indigenous communities, often referred to as *comunas* and inhabited by so-called *comuneros*. Such processes of urban expansion also had political and administrative implications as the city grew beyond its own municipal boundaries. This is particularly evident in southern La Paz where the city expanded into rural territories administered by the municipality of Palca. Opposed political parties dominate La Paz's and Palca's municipal governments. The political party Movement without Fear (*Movimiento Sin Miedo*, MSM), which stands in opposition to the national government controlled by Evo Morales's Movement Towards Socialism (*Movimiento al Socialism*, MAS), was in charge of La Paz's municipal government when the fieldwork for this book was conducted in 2012 and 2016. During this period, the municipal government of Palca was headed by the MAS party. Aiming to capture the perspective of different indigenous peoples (including migrants and *comuneros* of different ethnic background, age, and gender) and national as well as local government authorities (affiliated with both MAS and MSM), it was decided to work within a specific part of the city – that is, the neighbourhoods of Chasquipampa and Ovejuyo – where it was possible to approach this diverse portfolio of social actors.

The field sites: Chasquipampa and Ovejuyo

Chasquipampa and Ovejuyo are two neighbourhoods situated in the south-eastern part of La Paz. The majority of residents self-identify as indigenous, with 80 per cent being of indigenous migrant background and around 20 per cent claiming to be of *comunero* origin. The neighbourhoods are situated on politically contested terrain, in the heartland of the municipal-boundary conflict between La Paz and Palca. In order to visit these neighbourhoods, it is necessary first to take a minibus or cable car from La Paz's central neighbourhood of San Jorge. From there, one drives through the affluent districts of Obrajes and Calacoto. Once in Calacoto, it is necessary to change to another minibus which drives first through the district of Cota Cota before arriving in Chasquipampa and Ovejuyo.

Comuneros represent the original residents of Chasquipampa and Ovejuyo. Throughout the colonial and republican periods, they worked as landless peasants under semi-feudal conditions on *haciendas* (Espinoza 2004). Following land reforms in 1953, the *hacienda* system was abolished. *Comuneros* were organised within peasant unions and granted individual and collective land titles. The municipality of Palca, together with the national government, administered the land redistribution for *comuneros* (Espinoza 2004). By the late 1980s, Chasquipampa and Ovejuyo were becoming increasingly affected by the physical expansion of the city of La Paz. Seeking to profit from urban expansion, some *comuneros* in the area sold parts of their plots to land speculators or to newly arriving indigenous migrants who came mainly from rural Aymara communities near the city of Achacachi in Bolivia's Lake Titicaca region. These processes of land speculation and land use transformation were not unique to Chasquipampa and Ovejuyo but have been observed in other Bolivian cities (Goldstein 2013).

From the 1990s onwards, Chasquipampa and Ovejuyo transformed from rural areas into urban neighbourhoods. Because of the rapid urbanisation taking place outside its formal municipal boundaries and with an interest in expanding its local tax base, the municipality of La Paz redefined its municipal boundaries in 1995 by means of municipal law 1669. Chasquipampa and Ovejuyo, along with other areas affected by urban expansion, were now considered part of La Paz's municipal territory (Arbona and Kohl 2004). La Paz's legal claim to newer peripheral urban settlements was neither fully legal, nor was it uncontested. The municipality of Palca, currently with support from the departmental government of La Paz and the national government, continues to claim administrative authority over these neighbourhoods.

Organisational life within Chasquipampa and Ovejuyo is complex not only because of the presence of two different, and politically opposed, local governments but also because indigenous residents are themselves organised in a diverse set of community-based organisations (CBOs). *Comuneros* remain organised within peasant associations which mainly negotiate with Palca. By contrast, indigenous migrants have mainly formed neighbourhood associations which negotiate with La Paz to secure tenure rights and access to basic services. As will be demonstrated in Chapters 5 and 7, this organisational divide between *comuneros* and migrants is by no means clear-cut but is characterised by important overlaps, especially with *comuneros* trying to manoeuvre between rural peasant and urban neighbourhood organisational life. In addition, migrants and *comuneros* tend to be members of multiple CBOs, including folklore clubs, school clubs, and urban youth tribes, each of which follows a different agenda.

The contextual background information outlined earlier is important for understanding why and how different indigenous residents and national and local government officials understand and address urban indigeneity and interpret indigenous rights to the city. Chapters 5–7 will explore these issues in further detail but, before doing so, let us shift attention to Quito, Ecuador – the other case study examined in this book.

Indigenous in Quito: out of the Virgin's sight

In 2012, 2.2 million people lived in the metropolitan district of Quito and 1.6 million lived within Quito's urban and suburban areas (DMQ 2012b). According to 2010 census data, 10 per cent of Ecuador's current indigenous population, and approximately 25 per cent of its urban indigenous population, reside in Quito (INEC 2014). At the city level, seven per cent of Quito's population, approximately 150,000 people, self-identify as indigenous. Hence, in terms of size and proportion, Quito has fewer indigenous inhabitants than La Paz.

The relatively small size of Quito's indigenous population was, however, questioned by most indigenous residents approached in this study who generally pointed out that more people are likely to be indigenous than the census reports. Research participants generally explained gaps between reported and actual number of indigenous residents by highlighting that the people conducting the census fail to ask people whether they self-identify as belonging to a specific indigenous or ethnic minority group. Others said that they would not identify themselves as indigenous because of fear of discrimination (see also chapter 5). This is not a surprising reaction in a city where problems of racism remain paramount. For example, according to a survey undertaken by the municipal government of Quito, 12 per cent of the city's population state that they do not want indigenous people as their next-door neighbours. In this sense, being indigenous in Quito is closely related to problems of discrimination and invisibility (see Chapter 5). The problem of invisibility or of 'being out of sight' was explained by Franklyn, an indigenous activist and community teacher:

> The history of our city can best be understood from the perspective of the Panecillo [hill in the centre of Quito, see Figure 4.6]. Long before

Figure 4.6 View of Quito and the Panecillo hill.

the Spanish arrived here, this was a settlement and spiritual site inhabited by our ancestors. These included the Karas and Kitus, from whom our present city derives its name of Quito. The colonisers destroyed this settlement on the Panecillo hill, and north of the hill they constructed their own town. On the right side of their town, and in the shadows of the Panecillo, they forced the remaining *indios* into tiny houses. Here life was hard and miserable and, honestly speaking, it remains like that for the poor souls who still live here today. This part of the city is where we are now, San Roque, one of the areas with a particularly large indigenous population. Over the years Quito grew and more indigenous people came to live here in San Roque and in other parts of the city. But we are out of sight of the monument which supposedly guards this city from the top of the Panecillo. When I refer to the monument, I mean the *Virgin de Quito* which was constructed in the 1970s by a Spanish artist. You see, this Spaniard constructed a Virgin who looks straight up north. From this perspective, the Virgin can see the colonial centre and also the business districts and neighbourhoods of the rich *mestizos* who live in North Quito. San Roque, where we are now, is out of the Virgin's sight. The Virgin also turns her back towards South Quito and cannot look over the hills at the far north-end of the city. But it is precisely in these parts where most of us – the children of the Karas and Kitus – now live. We are out of the Virgin's sight, we are living in a city which is here not for us but only for the people that colonised and continue to colonise this place. I tell this story to all my indigenous pupils and now I tell it to you so that you can share it with more people and, hopefully, one day everyone in this city will be able to look one another in the eye.

(Interview, 26 April 2013)

Franklyn offers a captivating account of urban coloniality as experienced in everyday life by urban indigenous residents and a succinct description of historically established ethno-racial and spatial divisions in Quito. The city which is now called Quito was already populated prior to the Spanish conquest. According to Lozano (1991), human activities within the area can be traced back to 4000 BC when the Kitu people are believed to have lived in the area. The Kitus were later conquered by the Karas, Shiris, and Incas. In 1542, the Spanish colonisers conquered the Inca settlement and built their own city (Zaaijer 1991). As in La Paz, the colonisers constructed a segregated urban space in which 'white' *criollos* inhabited a central area planned according to a grid with a main square – now called *Plaza Grande* – at its centre. The *Plaza Grande* remains not only Quito's but also Ecuador's political centre as both the presidential palace and the city hall of Quito are located there.

Indigenous peoples were relocated into the local 'Indian' quarter of San Roque, which is situated a few hundred yards up the mountain on the north-west of today's colonial city centre (Salomon 1988). The indigenous peoples residing in this area mainly worked as servants for the *criollo* elite living on the lower slopes of the valley in the colonial city. Early urban indigenous residents also traded agricultural products at local markets in San Roque. In the colonial period, many of Quito's indigenous residents were impoverished and lacked access to adequate food, water, and basic services (Milton 2005). It is unclear how many indigenous people lived in Quito during the early colonial period. However, data from 1784 show that, of Quito's approximately 24,000 residents, 18,000 were *criollos*, 4,500 were *mestizos*, and only 1,500 were indigenous (Salmoral 1994).

During the late colonial and early republican periods, the urban indigenous population of Quito is believed to have declined (Milton 2005). Three explanations are normally offered (Salmoral 1994; Salomon 1988). First, hard labour conditions and lack of access to food and healthcare led to an increase in disease and death among indigenous peoples. Second, indigenous peoples interbred with *criollos* and gave birth to *mestizos*. Third, *mestizaje* occurred not only through biological interbreeding but also through cholofication, generally defined as changes in indigenous peoples' cultural and socio-economic practices within urban settings.

While Quito's indigenous population declined, the *mestizo* population continued to rise, leading to urban expansion and the emergence of wealthier middle- and upper-class neighbourhoods in areas north of the city centre such as La Floresta, Miraflores, La Colon, or Guapulo. Throughout the 20th century, affluent city residents increasingly relocated to lower-lying valleys east of Quito such as Cumbaya or Tumbaco. As in La Paz, however, indigenous urban growth and expansion did not begin to accelerate in Quito until the mid-20th century. It was then that rural peasants, predominantly of Kichwa origin, migrated to the city in search of work or access to education (Zaaijer 1991). Pull factors stimulating the migration to Quito included failed land reforms which made it difficult for highland indigenous peoples to make a living as peasants. The influx of indigenous and other migrants led to a process of unplanned urban growth and expansion. Incoming indigenous migrants initially settled in areas north-east of the historical city centre, namely within the already existing indigenous neighbourhood of San Roque. Here, they often rented rooms in densely occupied rundown colonial houses which lacked access to water and sanitation services (Espin 2012). In search of their own homes, indigenous migrants increasingly relocated to Quito's growing peripheral neighbourhoods, such as Chillogallo, Carcelen, Guanmani, or Quitumbe in the southern part of the city, or to suburban areas, such as Calderon in the far northern periphery.

Within their new neighbourhoods, indigenous migrants either rented houses or built their own homes on land which they either occupied illegally or bought from land speculators. As in La Paz, Quito's indigenous migrant population often lived – and continues to live – within areas characterised by lack of access to physical infrastructure services. For example, in 2012 the municipality of Quito reported that approximately 45,000 houses, homes to 180,000 residents, in predominantly indigenous neighbourhoods such as Quitumbe and Calderon lacked not only tenure rights but also access to basic public services such as electricity and water (DMQ 2012c). The report also highlights a shortage in social infrastructure such as schools and hospitals in these more indigenous neighbourhoods.

Unlike in La Paz, in Quito indigenous migrants often settled in different neighbourhoods with their families, without necessarily reproducing their rural indigenous communities within the city (Espin 2012). Instead, indigenous communities were recreated through collective organisation around their places of work (Kingman 2012). Today, the majority of Quito's indigenous migrants engage in commercial activities. They often work as market or street vendors, artisans, food carriers, or vegetable peelers in one of Quito's large markets such as San Roque situated in the Centro Historico, the Mayorista market near the Chillogallo neighbourhood, or the markets of the suburb Calderon. Indigenous migrants from the same community of origin often organise themselves in commercial associations and trade unions.

Around 140,000 of Quito's approximately 150,000-strong indigenous population are migrants of predominantly Kichwa descent. The remainder are so-called indigenous *comuneros*, who live in legally recognised communes. Jeremy Rayner (2017) emphasises that Quito's communes existed long before the colonial conquest. With the arrival of the colonisers, the indigenous peasants residing in these areas were integrated into the *encomienda* (the system whereby a Spanish colonist was originally granted a tract of land or a village together with its indigenous inhabitants) and later into the *hacienda* system. Following the 1937 Law of Communes, the territories of these communes were registered by Ecuador's Ministry of Agriculture and *comuneros* were entitled to govern themselves through autonomous *cabildos* (councils). Ecuador's current constitution continues to recognise the autonomous status of communes and obliges government authorities or other external stakeholders to consult *cabildos* before making any intervention in their territory.

The metropolitan district of Quito currently contains 49 legally recognised indigenous communes (see Figure 4.7) which are home to approximately 10,000 *comuneros*. The communes have been affected by processes of urban expansion occurring in the city since the second half of the 20th century. By 2012, 24 communes were already part of urban or suburban Quito. As in La Paz's rural-urban fringe, land was often illegally subdivided by *comuneros*, land speculators, or municipal authorities. As a consequence, urbanised communes are now inhabited by a diversity of indigenous and non-indigenous residents. They also represent important sites for business, industrial, and transport activities. This is especially the case for the communes of Oyambaro and Agila, where Quito's new airport is situated. In a context where communes are increasingly affected by urbanisation, indigenous leaders from various *cabildos* have started to organise themselves as a single movement representing the Kitu and Kara peoples – the living descendants of the precolonial civilisations which have long inhabited the territory of today's Quito (Rayner 2017).

In contrast to La Paz, where different municipal governments seek to exercise political control within one city, metropolitan Quito is governed by

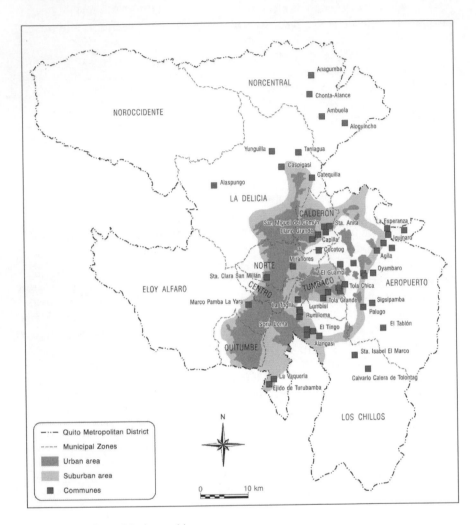

Figure 4.7 Map of Quito and its communes.

one municipal government and a variety of small-scale autonomous indigenous *cabildos* which represent the 49 communes within the metropolitan district. *Cabildo* leaders interviewed in Quito generally said that they do not associate themselves with any political parties and engage in politics purely in the interests of the residents of their *comuneros*. During initial fieldwork in 2013, Quito's municipal government was led by Augusto Barrera of the left-wing Alianza PAIS party, which at that time headed the national government. In office since 2009, Barrera was voted out in municipal elections in 2014. His place was taken by Mauricio Rodas of the opposition party *Sociedad Unida Más Acción* (SUMA).

Research with urban indigenous migrants and comuneros in Quito

Instead of conducting fieldwork within one specific neighbourhood, it was decided to work with specific organisations that represent indigenous residents living in different parts of Quito. Reflecting on the composition of Quito's indigenous population, it was decided to conduct fieldwork predominantly with indigenous migrants who represented the city's urban indigenous majority. Nevertheless, selected interviews have been conducted with *comuneros* residing in areas affected by urban expansion, including the communes of Agila, Llano Grande, Lumbisi, Oyambaro, and San Miguel del Comun.

More in-depth research was conducted with a group of indigenous migrants who originally came to Quito from the rural parish Tigua, situated in Ecuador's Cotopaxi province. Overall, some 7,000 Tiguan households live in various neighbourhoods of Quito. Tiguans are renowned in Ecuador and internationally for their colourful paintings depicting scenes of rural communal life. In Quito, approximately 300 Tiguan migrant households are engaged in the production and sale of paintings. This group of Tiguan artists mainly sell their paintings at the weekend market in the El Ejido Park in Quito's tourist district, Mariscal Sucre. The remaining Tiguans work as market vendors or food carriers in Quito's San Roque and Mayorista markets.

Like most of Quito's indigenous migrants (Kingman 2012), Tiguans have created their urban indigenous communities and organisational networks around their workplaces. The group of Tiguans approached in this research was organised predominantly within the market vendor associations AECT-Q (Tiguan Association of Carriers and Commercial Vendors Residing in Quito) or AVIC-Q (Association of Indigenous Vendors from Cotopaxi Residing in Quito). These associations met close to their members' workplaces. For example, the AECT-Q convened its meetings on the premises of the indigenous weekend school Chaquiñán College, situated near the central San Roque market. In addition to serving as a meeting ground for the AECT-Q, this school provided weekend education for 85 indigenous migrants from Tigua aged between 18 and 55. Like the wider Tiguan community, most of these students worked as market vendors and lived in peripheral neighbourhoods of the city. The Chaquiñán College therefore served as an ideal place to approach indigenous leaders and members of the wider Tiguan migrant community living in Quito.

Conclusions

This chapter has offered a historical overview of processes of indigenous urbanisation in La Paz and Quito. La Paz and Quito certainly share similarities – both are seats of their countries' national governments, and both experienced rapid urbanisation and urban expansion during the second half of the 20th century. In both, urbanisation and urban expansion coincided with processes of ethno-racial diversification and an increasing urban indigenous presence. The urban indigenous peoples of both cities are diverse, representing migrants from different parts of the country but also *comuneros* whose territories were absorbed during urban expansion. Urban indigenous peoples in both cities generally experience poorer socio-economic living conditions than non-indigenous residents.

There are, however, important differences between these two cities. First, more government actors with distinct political affiliations influence urban policy and planning in La Paz – a city where the municipalities of La Paz and Palca compete for political control – than in Quito – a metropolitan district governed by a municipal government but also by a set of autonomous indigenous *cabildos*. Second, even though smaller in terms of its overall urban population, La Paz is home to a significantly larger indigenous population than Quito. Third, the organisational structure of indigenous peoples differs between the two cities. In La Paz, indigenous peoples, migrants, and *comuneros* alike often live in peripheral neighbourhoods together with members of their communities of origin. Within these neighbourhoods, migrants and *comuneros* engage in different residential organisations. By contrast, Quito's indigenous population, particularly migrants who represent the city's indigenous majority, live more dispersed across the city and organise around their places of work.

The aforementioned similarities and differences were carefully considered in the design of the variation-finding comparative approach which underpins this study. As already outlined in Chapter 1, this approach assesses variations in the findings *between* La Paz and Quito mainly in relation to the unique processes and factors *within* each city. Such differences in internal processes have rarely been captured in previous comparative research, which predominantly studied the incorporation and implementation of indigenous rights at the national and local levels in relation to global and regional political trends. The intention of the following chapters of this book, therefore, is to contribute to existing comparative research in the field of Latin American indigenous studies by showing how country- and city-specific processes lead to slightly different understandings of indigeneity and attempts to address and translate indigenous rights to the city in La Paz and Quito.

Part II

Experiences from La Paz, Bolivia, and Quito, Ecuador

5 Urban indigeneity as lived experience

Introduction

The indigenous right to the city, despite being recognised within Bolivia's and Ecuador's constitutions, should not automatically be viewed as a purely legal category. Instead, the indigenous right to the city mainly represents a demand raised by ordinary indigenous residents to plan and design cities according to their specific interests and needs. Before exploring how the indigenous right to the city is taken up, addressed and translated by different actors into policy and planning practice, it is therefore important to identify how various urban indigenous residents themselves understand indigeneity and what are their specific interests and needs.

This chapter explores the aforementioned questions in the context of the selected case-study cities of La Paz and Quito. It offers an account of the lived experience of urban indigeneity. It finds that indigenous residents in both cities share a common desire to move ahead economically. In addition, indigenous residents in both cities articulate specific interests and demands in relation to land. Depending not only on their background as migrants or *comuneros* but also on their gender, age, and social status, they tend to associate land with distinct opportunities to access a portfolio of assets and varying claims to rights. Hence, indigeneity means different things to different residents and, likewise, there exist multiple and, at times, conflicting articulations of the indigenous right to the city.

Drawing on the asset-accumulation framework outlined in Chapter 3 and on the logic of a variation-finding comparison, the first section of this chapter explores a variety of articulations of indigeneity and associated interests, demands, and rights-based claims within the neighbourhoods of Chasquipampa and Ovejuyo in La Paz. The second section discusses these issues for Quito's indigenous migrant and *comunero* population. The final section compares the findings from both cities.

Urban indigeneity in La Paz's Chasquipampa and Ovejuyo neighbourhoods

Most of Chasquipampa's and Ovejuyo's residents self-identified as indigenous with Aymara origin. But these residents rarely match the specific criteria that are used for the definition of 'indigenous original peasants' in Bolivia's 2009 constitution and associated legislation. Bolivian intellectual Raul Prada explained why this is the case:

> In neighbourhoods like these, no one can claim to be a true indigenous person because the people that live there are no longer the ancestral owners of the territory. They are not organised in the traditional *ayllu* structure. They do not have an organic relationship to nature and mother earth. This is all impossible in the city. They are not what the constitution considers indigenous original peasants.
>
> (Interview, 30 January 2013)

Indeed, *comuneros* – residents who inhabited these areas prior to the arrival of the city – lost most of their ancestral land and now represent a minority. Similarly, indigenous peoples who migrated and settled in Chasquipampa and Ovejuyo left their ancestral land behind. In this sense, Prada is correct: Neither *comuneros* nor migrants represent what the constitution refers to as 'indigenous original peasants.' But why then do so many residents in Chasquipampa and Ovejuyo nevertheless self-identify as indigenous? What does indigeneity mean to people in these neighbourhoods? The indigenous migrant Pascual provides some answers to these questions:

> What they write about us indigenous people in the law is not reality. The times are changing and so are we. We learned and experienced new things. This makes it difficult for us to return to something we did in the past. It's as if I wanted my son to return to our old land at Lake Titicaca and live a backward life. He grew up here, went to school in the city, and knows his way around in this place. This doesn't mean that he has forgotten where he comes from. He speaks Aymara with us at home and with friends in his rap band. He stays in touch with the community at Lake Titicaca and practises his traditions at our neighbourhood festivals. You see, all of us are part of two worlds. We live in the city, but our cultural background of the past continues to shape our present. We were indigenous in our place of origin and will stay indigenous in this place.
>
> (Interview, 1 November 2012)

Relocation from ancestral territories, according to Pascual, should not simply be associated with a loss of one's indigenous identity and associated traditions and practices. Instead, migrants preserve, revitalise, and adjust their identity within their new urban homes. Carlos, a *comunero* and member of an indigenous peasant union, makes a similar point:

> This land belongs to my indigenous brothers and sisters. It has been taken away from us by corrupt land speculators. We might not get it

back. All we want is to live here and preserve our traditions as much as we can in this context.

(Interview, 30 January 2013)

In contrast to static understandings of indigeneity which associate indigenous peoples with rurality, tradition, and backwardness but not with modern cities (see Chapter 2), most indigenous residents in Chasquipampa and Ovejuyo aspire to combine a modern urban life with their traditions. This is evident in Pascual's description of his son who practises the Aymara language through rap music. Roberto, a young *comunero* from Ovejuyo, offers another explanation:

I work my land here. I grow my potatoes and herd some animals. I am engaged in the indigenous peasant union. I dance during the festival. At the same time, I do my job as a construction worker in El Alto. This helps me to pay the bills for my house. I want to go to university and create my own business. This is how a lot of people live their lives or, better, how we live here as indigenous peoples of the city.

(Interview, 24 January 2013)

Similar aspirations of mixing indigenous traditions with modern urban amenities were articulated by most indigenous residents approached in Chasquipampa and Ovejuyo. In particular, residents referred to the important role of land when defining their indigenous identity and associated interests, demands, and rights-based claims. Julia, a young migrant from Ovejuyo, described the importance of land like this: 'For us indigenous peoples, land is much more than just soil. It is the territory on which we live, work, enjoy our free time, and practise our culture' (Interview, 15 November 2016). Understood as such, land is associated with a portfolio of different assets. As will be illustrated later, though, land also has different connotations for distinct indigenous residents such as migrants, *comuneros*, men, women, youngsters, or the elderly.

Land as a physical asset

Insecurity of tenure is one of the biggest problems in the neighbourhoods. Table 5.1, for example, illustrates how, for a focus group composed of eight Aymara women, absence of secure tenure represents the main problem, followed by insufficient street lighting, bad roads, insecurity, and alcoholism. These women highlighted that the main causes for insecure tenure include municipal boundary conflicts between La Paz and Palca (see also Chapters 4 and 6), prevailing corruption, and personal conflicts with neighbours (see also Figure 5.1). Immediate solutions to resolving tenure insecurity include the permanent physical occupation of one's plot. Otherwise, longer-term solutions include opening a court case and seeking advice from a lawyer.

Table 5.1 The main problems in Ovejuyo[1]

List of problems (1)	Ranking (2)	Prioritisation (3)
Insecure land titles	3 + 3 + 3 + 3 + 3 + 2 + 3 + 2 = 22	1
Insufficient street lighting	2 + 1 + 3 + 3 + 2 + 3 + 2 + 1 = 18	2
Bad roads	3 + 3 + 3 + 3 + 1 + 2 + 1 + 3 = 17	3
Insecurity in the area	3 + 3 + 3 + 2 + 1 + 3 + 2 + 2 = 19	2
Bad consumption of alcohol by neighbours	2 + 1 + 1 + 2 + 2 + 3 + 3 + 2 = 14	5
Lack of money	2 + 2 + 2 + 2 + 1 + 2 + 1 + 2 = 14	5
Bad service by bus drivers	1 + 2 + 2 + 2 + 1 + 2 + 2 + 2 = 14	4
Garbage rarely collected	1 + 1 + 1 + 2 + 1 + 2 + 1 + 2 = 11	6

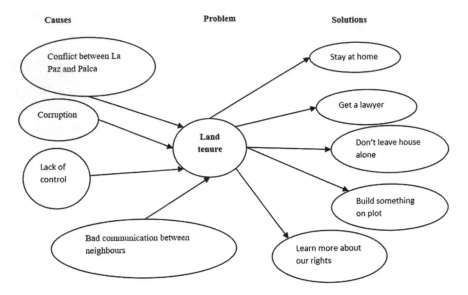

Figure 5.1 Causes and solutions to the land tenure problem in Ovejuyo.

Maria, an indigenous migrant, provides a detailed account of how residents cope with tenure insecurity:

> Most residents in the neighbourhood have a land title that was authorised by either La Paz or Palca. The two municipalities sometimes authorise land titles for the same plot of land. My family is affected by this problem of double tenure. In 1994 we bought a plot of land in Ovejuyo. With the help of friends who already lived in the area, we constructed our own house. In 1998 we started the complex bureaucratic journey of applying for a land title. We received some support from the local neighbourhood association. It took us two years to fulfil all the bureaucratic formalities to eventually receive a land title from La Paz. In 2002 a person who

identified himself as *ex-comunero* from Ovejuyo knocked on our door and claimed that the plot belonged to him. He proved that he owned this land by showing us an authorised land title from Palca. We closed our door so that he could not enter. From that moment onward, to avoid that the *comunero* takes over what he considers to be his plot, I make sure that our house is always occupied. The *comunero* returned to our house a few more times but I prevented him from taking over our plot.

(Personal communication, 7 December 2012)

Alongside the shared problem of insecure tenure, migrants and *comuneros* associate land with different aspirations and opportunities to access, accumulate, and maintain other physical assets. For migrants as well as for *comuneros* with an urban home, an individual land title is considered crucial as it is the precondition for accessing water, sanitation, electricity, and paved roads – services which are mainly provided by the municipality of La Paz (see Chapter 6).

In contrast, *comuneros* who own some of the few remaining plots of agricultural land consider themselves to be the ancestral owners of the lands that comprise the two neighbourhoods. These residents have, at times, perceived ongoing processes of urban expansion as a threat associated with the loss of their land. In addition, as pointed out by Francisco – a *comunero* who has an urban home in Chasquipampa and a plot of agricultural land in Ovejuyo – some *comuneros* also aspire to reclaim tenure rights over lost territories:

This land was agricultural land and it belonged to us. We want the state to recognise this land again as our collective territories. We want to manage our land according to our own ancestral principles and traditions. We also do not want to pay charges for services from La Paz. We do not need the municipal government of La Paz interfering here.

(Interview, 19 November 2012)

Land is, hence, associated with very different aspirations, ranging from leading a modern urban life to preserving a rural peasant lifestyle. Such interests are often articulated through claims to individual or collective rights to land, with the latter being mainly highlighted by *comuneros*. The underlying reason for *comuneros*' collective rights-based claims, however, should not automatically be conflated with interests in preserving traditional agricultural practices and preventing further urbanisation.

Land as a financial, natural, and productive asset

Pedro, a migrant from Chasquipampa, explained that many *comuneros* in the neighbourhood attach a specific productive and financial value to their land:

They refer to their land as rural and as collective mainly to pay less tax. Being rural means you are with the municipal government of Palca which hardly charges tax. Some *comuneros* own big extensions of land. Imagine how much money they save by not paying taxes! Some *comuneros* might indeed use their land for agricultural activities but most of them use it as future investment. You should know that the *comuneros* have power in this area. Everyone who owns large plots of land has power. At the moment, land might not be that expensive in Chasquipampa, but it is getting more so over time. In the future, a square metre might be worth US$100. If you own a big plot of land in this area, you can make real money in the future. Until this point you call this land rural, collective, ancestral, indigenous, or whatever. This allows you to keep it.

(Interview, 5 December 2012)

By registering plots as collectively owned rural land with the municipality of Palca, *comuneros* avoid paying urban public-service charges and gain legal permission to use the land for agriculture. It is for this reason that, within an otherwise dense urban neighbourhood, there exist small green spaces which are used for herding animals and cultivating vegetables (see Figure 5.2). Agricultural activities serve *comuneros* as a source of additional income. It is mainly women who are responsible for selling vegetables and meat products on one of La Paz's or El Alto's many street markets. More importantly, though, most *comuneros* use their agricultural land as

Figure 5.2 Land-protection strategies by residents in Chasquipampa.

future investment. The story of Francisco illustrates this trend. Francisco and his family used to own two large plots of agricultural land in Chasquipampa and Ovejuyo. In the early 1990s, Francisco and his brother decided to sell their land in Chasquipampa to incoming indigenous migrants. In 1998 Francisco married Diana, an Aymara woman from the rural community Huni which is situated south of Ovejuyo and belongs to the municipality of Palca. Diana and her family owned another plot of agricultural land in Huni. In 2013, Francisco decided to sell some of his land in Ovejuyo to people interested in investing and constructing residential homes. Returning to La Paz in 2016, Francisco and his family sold all of their land in Ovejuyo and moved to live in Huni. At that moment, the family still wanted to preserve their agricultural land in Huni, but Francisco warned that, with the city of La Paz expanding towards Huni, he was planning to sell this last remaining bit of family land for a good price.

Francisco's story is by no means unique. Many *comuneros* approached in Chasquipampa and Ovejuyo highlighted how holding on to one's collective, agricultural, or ancestral lands mainly serves the purpose of generating income through working the land, saving tax payments, or actively engaging in land speculation processes in a context of urban expansion. In fact, such relationships to land are considered an intimate part of the indigenous identity of *comuneros*. Alejandra, a young *comunera* from Ovejuyo, expressed this as follows (Interview, 5 December 2016): 'Like most people, we want to move ahead economically, and we do this by making autonomous decisions on how to manage and use our land. This is very much in line with indigenous peoples' quest for autonomy all over Bolivia.' Like Alejandra, *comuneros* approached in both neighbourhoods strategically draw on legislative discourse on autonomy and collective land management and governance (see also Chapter 6) in order, among other issues, to justify land subdivision, speculation practices, and the generation of economic profit.

Indigenous migrants living in the same neighbourhoods share similar aspirations. Many are also engaged in agricultural labour and in the trade of agricultural products on the city's markets. The case of the family of Jose – a migrant from Chasquipampa – illustrates this point. Jose, his wife, and four children originally came to Chasquipampa from a small village near Lake Titicaca. The family still returns frequently to this rural community to help the ageing grandparents during the potato harvest. After the harvest, the family returns to La Paz, bringing with them large quantities of potatoes and other goods, such as textiles and electronics bought cheaply at markets near the Bolivia-Peru border, which they sell for a higher price on street markets in the city. Such patterns of split-migration and economic practices based on kin relations are common for Aymara migrants in both neighbourhoods and seem to reflect a general trend in La Paz whereby indigenous migrants are reported increasingly to engage in national and transnational circuits of capital circulation, allowing the formation of a new urban indigenous middle class (see Tassi 2010; Tassi *et al.* 2013).

In addition to living off their land, indigenous migrants and *comuneros* generate a substantial part of their income by engaging in La Paz's urban economy. For example, while Francisco's wife Diana sells agricultural products at local markets, Francisco works as a bus driver in La Paz. Migrant Jose, in addition to the family trade business, also works as a builder in La Paz's and El Alto's construction sector. These income diversification strategies are not unique. According to statistical data provided by the municipality of La Paz (2010), 50 per cent of Chasquipampa's and Ovejuyo's population are registered as economically active within La Paz's urban labour market. Residents mainly engage in formal sector wage employment, the transport and construction sectors being most popular for men, domestic work and commercial activities being most popular for women. But some also reported that since the rise of Evo Morales as President of Bolivia, they had managed to get jobs in government institutions and banks that were historically confined to 'white' and *mestizo* elites. For example, among the indigenous residents whom I approached in La Paz, seven worked for the municipal government of La Paz, four for national government ministries, and three in banks and credit associations. For many of the residents approached, contemporary La Paz represents a boomtown with endless economic opportunities for indigenous peoples. Carola (Interview, 2 December 2016), a young *comunera*, explained this:

> In the Bolivia of Evo Morales, we indigenous peoples finally receive opportunities to achieve Vivir Bien in cities. We can find jobs in any sector, from business to government. Throughout the city there are jobs for us and this transforms us from poor, excluded *indios* to urban indigenous entrepreneurs.

But urban indigeneity does not only mean moving ahead economically. It also refers to preserving or reinventing cultural and ancestral traditions.

Land as a cultural asset

Chasquipampa's and Ovejuyo's indigenous residents, migrants and *comuneros* alike, attach a specific cultural meaning to vacant land and public spaces in their neighbourhoods. This becomes evident during the festival *Fiesta de la Virgen de Merced* which takes place every September for three days and nights. Of La Paz's 387 annual folkloric festivals, the *Fiesta de la Virgen de Merced* is the second largest after the *Fiesta del Gran Poder*. During the festival the main road of Chasquipampa is taken over by thousands of Aymara residents. Participants are mainly *comuneros* and migrants from local neighbourhoods. On the main road of the neighbourhood various folkloric associations entertain the cheering crowd with presentations of the *Morenada*, a traditional Andean dance (see Figure 5.3). In the side streets visitors can purchase traditional snacks and beer.

Figure 5.3 Scene from the *Fiesta de la Virgen de Merced* in Chasquipampa.

The origins of the *Fiesta de la Virgen de Merced* date back to colonial times when today's neighbourhood of Chasquipampa was still part of a rural *hacienda* (see also Espinoza 2004). According to oral histories collected from *comuneros*, ancestors saw a virgin saint on the fields of the *hacienda*. The virgin saint, whom they called *La Mamita Merced*, promised a good harvest which indeed transpired in the following months. To commemorate this occasion, the ancestors constructed a chapel for the virgin saint which now lies on the main street of Chasquipampa. In front of this chapel, the ancestors of today's *comuneros* organised a festival every September to honour the virgin saint and to prepare for a good planting and harvest season. To ensure that the planting season is followed by a productive harvest, peasants relied on festive rituals such as the *ch'alla*. During the *ch'alla*, people donate parts of a drink, normally alcohol, to mother earth (in Aymara: *pachamama*) and drink the remainder. Today's *comuneros* keep this tradition alive.

The *Fiesta de la Virgen de la Merced* is organised by folkloric associations presided over by *comuneros*. As many as 30 folkloric associations are involved in the organisation of the annual festival. But indigenous migrants are also involved in such associations. Migrant Jose explains this:

> I am glad to be part of the *folkloristas* [referring to the folkloric organisation]. With them I help organise the *Fiesta de la Virgen de Merced*. We used to have *fiestas* like this one in our communities on the *Altiplano*. Taking part in their *fiesta* [referring to *comuneros*] reminds me of the festivals in my own community.
>
> (Interview, 12 November 2012)

A similar explanation was provided by Pascual – another migrant from Chasquipampa:

> Back in my home in Achacachi, we celebrated well. We celebrated our animals and plants. To make them grow you have to share your drinks with the *pachamama*. At the *Fiesta de la Virgen de Merced* we do the same.
> (Interview, 1 November 2012)

The festival, hence, is not only an act of dancing and drinking. It also helps *comuneros* and migrants to revitalise their cultural practices. Diana, Francisco's wife, expressed this as follows: 'The *fiesta* brings the countryside to the city. During the festival we, the indigenous peoples of the neighbourhood, rule this place' (Interview, 19 November 2012). This confirms previous research by David Guss (2006) who considers folkloric festivals in La Paz as decolonial acts whereby indigenous peoples confidently take over territories which were historically configured as 'white' spaces.

Not everyone in the neighbourhoods, however, views the *Fiesta de la Virgen de Merced* as a positive example of the protection and revitalisation of indigenous culture. In fact, indigenous culture and the use of public space were understood differently by residents belonging to different gender and age groups. Most elderly men interpreted the festival and associated acts of excessive drinking as their highlight of the year. In contrast, most women perceived the annual festival as a problem as they associated excessive drinking during this event with an increase in violence in their neighbourhoods and in their own homes (see again Table 5.1). Further, while elderly residents mentioned the festival as an opportunity to revitalise their rural traditions in the city, younger residents, migrants, and *comuneros* alike consider the music and dances of the festivals as outdated and backward. Instead, they prefer to articulate their indigenous identity through Aymara rap music and graffiti (see Figure 5.4). According to Nancy Hornberger and Karl Swijnhart (2013), such practices which fuse the Aymara and Spanish languages constitute productive spaces for indigenous youth to revitalise their identity and position it within contemporary culture in a globalised world.

Indigenous residents also use other practices to preserve their indigenous traditions and culture. Roberto – *comunero* in Ovejuyo – explains this:

> Our neighbourhood of Ovejuyo is built on the ruins of *ayllus*. Where I live there always was an *ayllu*. Now this *ayllu* is part of the city of La Paz. The government says that *ayllus* only exist in the countryside. I say, the *ayllu* exists where indigenous people live. We live in the city, in the urban *ayllu*. Here we practise the ancient traditions of the *ayllu* – leadership rotation, *mitas* [collective work schemes], justice, festivals etc.
> (Interview, 24 January 2014)

As outlined in Chapter 2, the *ayllu* represents a traditional, precolonial form of governing and managing land in the Andean region.

Figure 5.4 Aymara rap performance in Ovejuyo.

Anthropological research highlights the fact that, particularly in rural areas of Bolivia, the *ayllu* structure remains intact in postcolonial times (Harris and Tandeter 1987; Platt 1982). Roberto goes beyond a ruralised understanding of the *ayllu* by explicitly referring to Ovejuyo as an urban *ayllu*. Similar to Roberto, indigenous residents in both neighbourhoods often referred to their neighbourhoods as *ayllus* to justify specific practices such as enforcing indigenous lynch mob justice.

In summary then, urban indigeneity in the neighbourhoods of Chasquipampa and Ovejuyo means different things to different residents, with most of them aspiring to move ahead economically and to fuse ancient rural traditions and practices with a modern life in the city. The diversity of indigenous identities is articulated through specific claims to land – an important asset associated with various opportunities to access other resources. It is also evident that indigenous residents in these neighbourhoods do not represent homogenous and harmonious communities. Instead, there exist important intra-group differences: For example, a small group of *comuneros* seek to preserve the remainders of their collectively owned land for agricultural purposes, while a majority want to sell this land for a good price and profit from the urbanisation processes affecting their neighbourhoods. All this suggests that indigenous identities and associated interests, demands, and rights-based claims are not only diverse but often in conflict with each other. Similar trends can also be noted in Quito.

Urban indigeneity among Quito's indigenous migrant and comunero population

Seven per cent (approximately 150,000) of Quito's population self-identified as indigenous in 2010 (INEC 2014). According to Oscar, an indigenous migrant from the organisation RUNAKUY, Quito's indigenous population is actually much larger, but many people hide their ethnic identity in order to protect themselves against the threat of racism and discrimination:

> Back in pre-colonial times we were all natives. In the mid-19th century almost half of Quito's population were indigenous. Currently we are less than seven percent. There were no massacres against indigenous peoples in Quito. There was no fever that wiped us out. We are seven per cent simply because many of us fear discrimination. In reality, there are a lot more of us but we are often too scared to show our indigenous identity in public.
>
> (Interview, 8 April 2013)

Luis, a *comunero* from the commune Agila, offers a vivid account on the everyday reality of racism and discrimination in Quito which indigenous residents tend to confront:

> My mother always brings my daughter Maria to her school in Tumbaco [suburban neighbourhood of Quito]. A couple of weeks ago the two of them embarked on their usual bus journey and the driver and conductor mocked my mother for her clothes and for her bad Spanish [Maria's native tongue is Kichwa and she speaks little Spanish]. They called her and my daughter *indias sucias* [dirty 'indians']. The conductor prevented them from taking a seat. He said to them 'Go all the way to the back so that no one else needs to smell and see you dirty indians.' We get treated like this a lot. Sometimes they don't let you onto the bus or into shopping malls. They make you feel like you don't belong to this city.
>
> (Interview, 6 June 2013)

Many indigenous migrants and comuneros approached in Quito shared similar stories, highlighting the problem of deeply entrenched patterns of racism. In fact, in focus groups, indigenous migrants from the Chaquiñán College highlighted discrimination against indigenous peoples as the most important problem they face in the city, followed by issues such as the increasing lack of community, insecurity, and drug dealing in the work environment (see Table 5.2). As regards to the last problem, it is important to mention that most indigenous migrants approached in this research worked at Quito's San Roque market, a renowned hotspot of crime, prostitution, and drug trafficking within the city (Kingman 2012).

When discussing the main reasons for discrimination, focus-group members mainly referred to racist attitudes by other urban residents but also to their physical appearance, lack of Spanish language skills, and rural background (see Figure 5.5). Hiding one's indigenous features was, therefore, seen as an easy solution. This can be achieved through wearing jeans and

Table 5.2 The main problems of indigenous migrants in Quito[2]

Problem	Ranking	Prioritisation
Bad treatment and discrimination against indigenous peoples	$3 + 3 + 3 + 3 + 3 + 3 + 3 + 3 = 24$	1
Lack of community	$2 + 3 + 3 + 2 + 3 + 3 + 3 + 3 = 22$	2
Insecurity	$3 + 2 + 3 + 1 + 2 + 3 + 2 + 2 = 18$	3
Lack of work	$3 + 3 + 2 + 1 + 1 + 3 + 3 + 1 = 17$	4
Drug dealers on the street	$3 + 2 + 3 + 1 + 2 + 2 + 1 + 2 = 16$	5
Being a young father or mother	$1 + 1 + 1 + 1 + 3 + 2 + 3 + 1 = 13$	6
Contamination	$1 + 1 + 2 + 1 + 2 + 2 + 1 + 2 = 12$	7

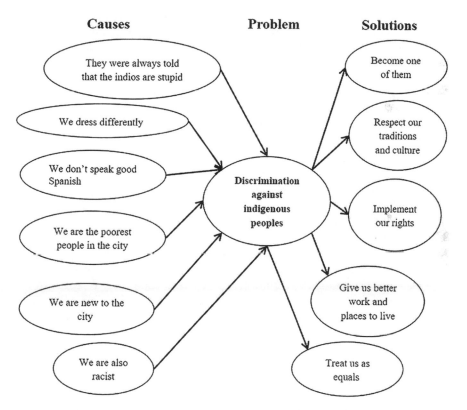

Figure 5.5 Causes and solutions to problems of racism and discrimination in Quito.

shirts instead of traditional ponchos or dresses, by taking Spanish lessons and speaking Spanish only in public spaces, or – especially in the case of men – by cutting one's long hair. According to Angelica, a student at the Chaquiñán College, such changes in appearance in public help Quito's indigenous residents to 'become one of them' (Interview, 13 May 2013). It means undergoing a process of 'whitening,' *mestizaje*, or cholofication (see Chapter 2).

In view of ongoing patterns of discrimination and associated responses by indigenous residents, it is unsurprising that only a small proportion of Quito's residents self-identifies as indigenous. At the same time, though, indigenous residents also mention that they often do not receive the option to self-identify around their ethnic identity. Rodrigo, member of the Tiguan Association of Carriers and Commercial Vendors Residing in Quito (AECT-Q) in San Roque, illustrates this point:

> No one told us that the census has a question about our ethnic background. When they knocked at our door they did not ask us whether we are indigenous or not. Maybe they just noted us down as *mestizos* because we do not look like one would imagine a stereotypical indigenous person. But we are indigenous peoples and it is sad that they make us invisible in the city.
>
> (Interview, 26 April 2013)

He argues that Ecuador's government strategically used the 2010 and previous censuses to portray indigenous peoples and associated social movements as minorities whose interests and demands were of marginal concern (see also Martinez Novo 2014 for a related discussion). It is for this reason that people responsible for collecting census data are not encouraged to ask people about their ethnic origin.

Despite being confronted by discrimination and strategic invisibilisation, most indigenous residents approached in Quito nevertheless say that they want to preserve specific indigenous traditions within the city. For example, when referring to solutions to the problem of discrimination, focus-group members also demanded respect for their traditions and cultures in an urban environment. As is the case in La Paz, though, they equally aspire to be treated as equals and to enjoy the amenities of an urban life – access to housing, public services, and economic and educational opportunities – whilst having their specific rights respected. When articulating such aspirations, indigenous residents also refer to the important role of land or urban space. But, depending on their status as *comuneros* or migrants as well as their age or position within their community, different residents associate land with opportunities to accumulate other assets which would allow them to lead rather distinct urban lives.

Indigeneity and the city: the case of comuneros

> We lived here before the city of Quito was built. Even though we are increasingly part of this city, the way we live here and manage our territory as an indigenous community makes us different from the rest. We try our best to protect Llano Grande from the evils of the city.
>
> (Interview, 30 April 2013)

Similar to the aforementioned testimony from Enrique, an elderly *co-munero* from the commune Llano Grande, other elderly *comuneros* mention that ancestral ties to their lands represent an important part of their indigenous identity as it allows them to preserve values of community. These *comuneros* know what life was like prior to the arrival of the city and they often aspire to return to their old peasant lifestyles. Many of them still cultivate potatoes or corn and preserve traditions such as the *minga* (collective work schemes), leadership rotation, indigenous justice, and the communal transfer of land from family to family. Jaime, another elderly *comunero* from San Miguel del Comun, explains this attachment to ancestral lands as follows:

> For us, being indigenous means to be recognised as owners of our ancestral territories. On our territories we could maintain our traditions. The way we worked and inhabited our territories defined our indigenous community. Our food used to come from our territories. Our festivals and dances honoured different seasons which affected our territories. Now, with the city coming closer, our territories are under threat.
>
> (Interview, 25 April 2013)

For elderly *comuneros*, then, land is associated with opportunities to access productive (cultivation of crops), cultural (festivals), social (community), and political (territorial autonomy and self-governance) assets, which, together, define and shape their indigenous identity. The urbanisation of their ancestral land is consequently associated with a loss of their indigenous identity and of their territorial autonomy.

Urbanisation changed the demographic composition of communes. Angelica, a young *comunera* from the commune Lumbisi, stated, for example, that, in the past, communal lands belonged to only 12 *comunero* families. With some *comuneros* subdividing and selling their land, the population of the area increased significantly and, in present-day Lumbisi, there live more than 2,000 inhabitants with different socio-economic and ethno-racial backgrounds. The 12 families representing indigenous *comuneros* are now a minority (Personal communication with Angelica, 15 April 2016). *Comuneros* from other communes outlined similar tendencies.

In 1993 Quito's 49 communes were officially integrated into the metropolitan district of Quito (DMQ 2012a). From then on, Quito's municipal government involved itself in the territorial management of communes, initially without respecting communes as semi-autonomous jurisdictions (for a detailed discussion, see Chapter 6). Municipal authorities issued newly arriving residents with individual land titles and provided them with access to services. Such activities were often undertaken without the consent of commune authorities who are organised in *cabildos*. Quito's municipal government also sought to include *comuneros* within their jurisdiction and offered to register their land with Quito and no longer with the Ministry of Agriculture. In exchange, the municipal government provided residents with access to water and electricity. Elderly *comuneros* recalled that they rejected such offers which they saw as a threat to their political autonomy. Enrique explains this as follows:

> Of course, we want water, electricity, roads, schools, and all this. We don't live in the jungle here but are part of the capital city of Ecuador. As *cabildos* we wanted to pay the municipality for services but this was not enough for them. You see, the problem is that they want us to turn into *barrios* [neighbourhoods]. This means it would be no longer our *cabildos* that make decisions here but the municipality of Quito. You know how this municipality works: they give us services but take our territories. You see how they give permission for people to build houses on our territories. Everything has a price but losing our rural territory and independence is too big of a price to pay.
>
> (Interview, 30 April 2013)

Not everyone in Quito's communes, though, perceives urbanisation and interventions by municipal authorities as a problem or threat. In fact, it is often *comuneros* themselves who contribute to the urbanisation of their own communes, by subdividing land and selling it to new residents. For some, then, land is associated with opportunities to score financial benefits. Further, contrary to elderly *comuneros*, younger *comuneros* rarely oppose interventions by the municipality of Quito but, instead, openly welcome them. Rebecca, a young *comunera* from Llano Grande, explains this as follows:

> Of course, I am an indigenous *comunera* but, unlike my grandfather or my father, I want to be part of this city. Our days as peasants are over. We want to lead a prosperous life. We may protect some of our community land and traditions, but when it comes to my own home I am with Quito. I registered with the municipality of Quito. Now they deliver electricity and water to my house and since then life is much better.
>
> (Interview, 16 April 2013)

Municipal government staff approached in Quito provided a very similar interpretation to Rebecca's, highlighting that most *comuneros* want to receive access to the public infrastructure and services provided by Quito. They did, however, also recognise ongoing patterns of resistance by elderly residents. To adequately address the diversity of interests of *comuneros*, the municipal government of Quito has in recent years established a co-governance unit which involves a diversity of indigenous residents from communes in processes of land management and territorial planning (see Chapter 6 for a detailed discussion).

In short, then, being indigenous in the city has multiple and sometimes contradictory meanings for *comuneros*. For the elderly it is associated with aspirations to regain autonomy over long-lost lands and to revitalise past traditions and agricultural practices. At the same time, younger *comuneros*, while not necessarily neglecting their indigenous roots, prefer to lead a modern urban life. For the latter, landownership is important, but it is mainly associated with opportunities to access services and generate financial benefits through subdividing and selling land.

Indigeneity and the city: the case of indigenous migrants

As highlighted in Chapter 4, the majority of Quito's indigenous peoples are migrants who came to the city from rural areas. Migrants often maintain strong ties with their rural homelands. Natalia, an elderly indigenous leader representing migrants from Chimborazo, illustrates this point (Interview, 18 April 2013): 'For me and for most of the members of my organisation, being indigenous means staying in touch with our rural community. It means knowing our language Kichwa and our traditions as indigenous peasants.' The importance of one's rural origin is also emphasised by younger indigenous second- and third-generation migrants who aim to maintain ties to their rural communities of origin, something which is often done by visiting grandparents or other family members during bank holiday weekends, attending annual festivals in their villages, or helping out with agricultural activities.

Despite such strong ties to their rural communities, indigenous migrants are predominantly engaged in making a living in the city. The majority of migrants approached in this research were working as market traders, street vendors, vegetable peelers, cooks, domestic workers, and folkloric art traders. They worked mainly at Quito's central San Roque market or at Quito's main market, the *Mercado Mayorista*, which is situated near the neighbourhood Chillogallo in the south of the city. Here, indigenous migrants often work six- to seven-day weeks, leaving their homes early in the morning and only returning late at night. With considerable time spent at work, markets represent important social spaces in which traditions of rural indigenous communities are preserved or revitalised. Daniel, a migrant from the Cotopaxi region who works at Quito's San Roque market, explains this:

> Here we all work next to each other. Of course we sell products but, more importantly, we are together with other people from our community. I work together with my daughter. The stall next to me is owned by one of my aunties. Opposite me work other people from my community in Cotopaxi.
>
> (Interview, 23 April 2013)

Similar trends could be noted for indigenous migrant market vendors who form part of the association AECT-Q. All of the members of this association originally came to Quito from the rural parish of Tigua. They sell their products on the same market lane, and, other than doing business, association members engage in daily banter; take care of each other's children; and, in the event that a conflict emerges, resolve it according to their own rules. For example, during one visit to the San Roque market, a member of the AECT-Q was accused of stealing money from another association member. To resolve the conflict, the vendors called the AECT-Q leader Juan who later punished the thief by telling him to pay back double the amount of money. According to Juan, such interventions resemble principles of indigenous justice that originate from his rural community of Tigua (Personal communication, 8 May 2013).

Within the San Roque market there also exist community centres in which indigenous migrants gather for association meetings, festivals, school lessons, or leisure activities. For the members of the AECT-Q, the Chaquiñán College was such a community space. Here, association members came together every year in June to celebrate the annual *Inti Reymi* – a festival of the summer solstice celebrated by Kichwa communities throughout the country (see Figure 5.6). Similar to the *Fiesta de la Virgen de Merced* in La Paz, the *Inti Reymi* helps to revitalise ancestral festivals in the urban context. During other times of the year, indigenous students of the college would use the community space to play sports or for hip-hop and punk concerts. As in La

Figure 5.6 Celebrating the *Inti Reymi* festival in the Chaquiñán College.

Paz, young indigenous migrants would perform songs in which they mix Kichwa language with Spanish and English slang.

During initial fieldwork in Quito in 2013, most migrants approached as part of this research still lived close to their work. Often they rented a small room in one of the large houses in the city's San Roque area that were owned by their market-vendor association. These houses were often overcrowded and lacked access to basic services. For example, approximately 200 indigenous migrants belonging to the AECT-Q lived in 50 small rooms in colonial-era houses in San Roque. For members of the AECT-Q, living in a community meant living in dire conditions and often was a decision made not out of choice but out of pure necessity. To improve their living conditions within the city, AECT-Q members generally expressed aspirations to live in a nicer house with better services and amenities (see Table 5.3).

To address their housing needs, some migrant families moved to other neighbourhoods in Quito. The case of Ricardo's family is illustrative. After three years of living in an overcrowded house managed by the AECT-Q, Ricardo's family moved to the peripheral neighbourhood of Guanmani in southern Quito in summer 2013, where they could afford to rent their own little flat which included a kitchen and bathroom. Even though Ricardo and his family now rented out a better place to live, they still faced problems:

> Of course, things improved for us. We now live in a nice flat but it is so far away from the San Roque market where we work. Sometimes it takes us three hours to get to San Roque from our flat in Guanmani. But this is not the most serious problem. We now live far away from our brothers and sisters [referring to members of his indigenous community that migrated to Quito] who live in the north, the centre, or elsewhere. Everyone moves to places which they can afford, but if we continue like this we will lose our indigenous community.
>
> (Interview, 11 May 2013)

Table 5.3 What would be the Quito of your dreams?[3]

What would make Quito a better city?	Number of focus-group members selecting this theme
Everyone should have a nice house to live in	16
Make Quito a safe city without crime and violence	13
Provide more work opportunities	12
Residents should respect each other/treat each other as equals	11
Create more green spaces and public spaces	9
Reduce contamination and create a clean city	5
Have a government that understands the needs of the people	4
No response	2
Total	72

When returning to Quito in 2015 and 2016, most migrants had relocated to rent their own apartments or houses in peripheral neighbourhoods. The importance of preserving community and a sensation of loss of community due to the increasing dispersal of indigenous migrants throughout the city was highlighted by many of those approached in this research. For example, when asked 'What does it mean to be indigenous in the city?' 14 out of 72 indigenous students from the Chaquiñán College said that they wanted to live in a community. Eleven of the 72 students were even more concrete – they wanted to be in a community in San Roque, not elsewhere in the city (see Table 5.4).

What is evident, then, is that indigenous migrants in Quito, like those approached in La Paz, want to enjoy a modern life in the city while preserving traditions and a sense of community. But they often struggle to do so. Oscar, an indigenous migrant and member of AECT-Q, expressed this as follows:

> Most of the people in Quito want to have a house, a car, and some money to spend on their family. We want this as well but, in addition, we want to build our community again in the city. Living, working, learning, celebrating, and experiencing life together in a community is the most important aspect for us indigenous migrants. It is difficult to combine these two worlds with most of us living further and further apart and with the local government destroying the last communal spaces we have.
>
> (Interview, 8 April 2013)

In this testimony, Oscar refers to the closure of the Chaquiñán College, which took place in 2013 as part of municipal government interventions in the San Roque area, which sought to promote the beautification, revitalisation, and regeneration of this district situated on the edge of Quito's colonial

Table 5.4 What would be a Quito for indigenous peoples?[4]

What do indigenous peoples want to live better in Quito?	*Number of focus-group members selecting this theme*
A city where we can live as a community	14
Not being discriminated against	13
To be allowed to stay as a community in San Roque	11
Be organised	9
Not being poor	7
Have better paid work	6
No responses	4
Change our identity/become *mestizos*	3
Have our rights respected	3
Respect nature	2
Total	72

city centre (see also Chapters 6 and 7). For members of the AECT-Q, though, urban regeneration represents a threat to the preservation of indigenous community. This was perhaps best expressed by Jose (Interview, 4 June 2013): 'We no longer have a space to be with each other in a community in San Roque. If they continue to take our spaces away, they will take our indigenous identity away.'

Conclusion

This chapter has explored what indigeneity means to those people who self-identify as indigenous in the selected case-study sites in La Paz and Quito. Indigenous residents in both cities can be broadly divided into *comuneros* and migrants. The characteristics of *comuneros* and migrants are, however, different in each city. La Paz's *comuneros* and migrants share the fact that they mainly live in the same peripheral neighbourhoods. In contrast, Quito's *comuneros*, agglomerated in 49 communes, rarely share their living spaces with indigenous migrants. Rather, migrants live dispersed across the city, an issue which is seen as problematic by most migrants who would prefer to live in a community.

Indigenous residents – whether *comuneros* or migrants – approached in both cities differ according to age, gender, and socio-economic status. Notwithstanding their diverse backgrounds, and regardless of whether they are migrants or *comuneros*, indigenous residents in Quito generally state that being indigenous in the city amounts to being a victim of discrimination. This is less the case in La Paz. A potential factor explaining this difference could be that La Paz – with about one-third of its population self-identifying as indigenous – is a far more 'indigenous' city than Quito, where fewer than seven per cent of residents consider themselves to be of indigenous descent. But political context also matters. As will be further demonstrated in the next chapter, in Ecuador national and local governments remain controlled by *mestizo* elites who are often guided by racist attitudes towards indigenous peoples. Bolivia, by contrast, having a president and other senior officials who self-identify as indigenous has, to a degree, broken down racist and discriminatory structures. That is why indigenous residents in La Paz no longer fear entering historically 'white' urban spaces such as universities, government buildings, or up-market neighbourhoods. Rather, indigenous residents now work within such institutions and increasingly inhabit middle- and upper-middle-class neighbourhoods.

The last point links to another common feature that characterises indigenous residents, migrants, and *comuneros* in both cities. All of them want to move ahead economically and, as will be demonstrated in the next chapter, Bolivia's and Ecuador's governments do indeed address these aspirations through a policy of redistribution. Indigenous residents in both cities also share the fact that they express their indigenous identity and associated interests, demands, and rights-based claims by reference to land. But, for different people, land is associated with distinct interests, demands, and opportunities to access rights and resources. An asset-accumulation framework helps to draw out this complexity and illustrates how land is at once associated with varying opportunities to access to cultural (e.g. festivals and collective land management), financial (e.g. money generated from reselling land), physical (e.g. access to water, electricity, or roads), social (e.g. meeting spaces), natural/productive (e.g. agricultural activities), or political (e.g. associated with rights for governance and autonomy) resources.

Within each city it is notable that indigenous residents associate land with opportunities to preserve traditions and practices which are considered stereotypical for a rural and authentic indigenous lifestyle (Field 1994). Elderly *comuneros* in La Paz and Quito particularly associate land with the preservation of a peasant lifestyle and with opportunities to cultivate agricultural goods. Similarly, *comuneros* seek to preserve collective ownership rights over their lands. They also want to manage their territories with autonomy and according to their own governance principles. Hence, processes of urbanisation and the growing influence of La Paz's and Quito's municipal governments are generally seen as a threat to their political autonomy. Associations of land as a collective good were highlighted not only by *comuneros* but also by indigenous migrants. Furthermore, as is seen throughout the *Fiesta de la Virgen de Merced* in La Paz and the *Inti Reymi* in Quito, migrants and *comuneros* both associate land with the opportunity to preserve and revitalise ancestral traditions.

In both cities, though, indigeneity refers not only to preserving tradition but to leading a modern life in the city. This is particularly the case for younger migrants and *comuneros* from both cities, who often fuse their indigenous traditions with Western culture, whether that is through Aymara hip-hop in Chasquipampa and Ovejuyo or through the fusion of Kichwa with Spanish or even English slang in Quito. In contrast to preserving ancestral landownership and management practices, some migrants and *comuneros* in Quito and La Paz also associate land with the accumulation of other assets which do not conform to a static, essentialised, and traditionalist understanding of indigeneity. They aspire to receive individual tenure rights from municipal governments in order to access urban infrastructure such as water, electricity, or roads. Like most urban residents then, they want to lead a modern life within the city and therefore seek to access universal rights and services which are provided to them by local authorities.

It is also important to note that indigenous demands for collective landownership or for the preservation of collective work, living and cultural spaces should not be automatically conflated with aspirations to preserve a traditional indigenous lifestyle. This is particularly evident in

La Paz, where some indigenous *comuneros* seek to preserve or regain access to collective land in order to later subdivide and sell it at a profit on the city's land speculation market. Hence, these *comuneros* associate land with monetary profit and often contribute to the destruction of their own 'indigenous communities.' Similar tendencies can be noted in Quito's communes.

Whereas previous research mainly focussed on selected cultural or economic demands of urban indigenous peoples (Albo *et al.* 1981; Guss 2006; Kingman 2012), the analysis presented here offers a more holistic understanding of what it means to be indigenous within the city. Within La Paz and Quito, indigenous residents articulate multiple indigenous identities. Hence, within these cities indigeneity is not a static but a dynamic social category which changes in meaning over time and space and means different things to different people. Likewise, indigenous residents express diverse, sometimes conflicting and contradictory, interests, needs and associated rights-based claims. This speaks against an understanding of a coherent set of indigenous rights to the city – articulations that resemble a collective struggle and joint outcry of common interests and demands. Instead, articulations of the indigenous right to the city vary within and between cities, making it difficult for government authorities to come up with a single coherent policy and planning agenda on this issue – a problem which will be discussed in further detail in the next chapter.

Notes

1 The information provided in this table was elaborated by the author and is the result of a listing and ranking of problems exercise undertaken with eight Aymara women (five migrants and three comuneras) in a community centre in Ovejuyo on 6 December 2012. Participants were asked to list any problem existing within their neighbourhood. They were then asked to rank each problem individually between 1 and 3, with 1 meaning not severe, 2 meaning severe, and 3 meaning very severe. The numbers were then added up. Based upon the ranking exercise, it was possible to determine which problems were seen as most serious. The problem with the highest ranking was marked 1 and was considered to be the most severe.

2 The information provided in this table was elaborated by the author and represents the result of a listing and ranking of problems exercise undertaken with eight indigenous migrants (five female and three male) in the Chaquiñán College on 8 May 2013. The listing and ranking of problems followed the same procedure as outlined in Table 5.1.

3 The information provided in this table was elaborated by the author and provides a summary of an icebreaker exercise undertaken during eight focus groups, which involved 72 Kichwa students from the Chaquiñán College.

4 The information provided here was elaborated by the author and provides a summary of another icebreaker exercise undertaken during eight focus groups with students from the Chaquiñán College.

6 Urban indigeneity in policy and planning practice

Introduction

Bolivia's and Ecuador's constitutions emphasise the importance of addressing indigenous rights, interests, and needs in cities. Most government authorities, policymakers, and planners approached in La Paz and Quito shared the view that it is the state that decides how and to what extent constitutional content on urban indigeneity is addressed in practice. As a senior official working for Ecuador's Ministry of Urban Development and Housing put it, 'The state is the guarantor of rights and this includes the right to the city' (Interview, 5 June 2013). This chapter, therefore, focusses on the practices of government authorities in La Paz and Quito. It reveals that, even in progressive settings such as Bolivia and Ecuador, the translation of indigenous rights to the city is by no means a straightforward process. Although indigenous rights are recognised through constitutional reforms, obstacles remain to the delivery of policies and planning interventions that are shaped to the interests of urban indigenous residents that were discussed in detail in the previous chapter. Gaps between legal rhetoric and practice may be explained through a variety of context-specific factors, including (1) prior constructions of indigeneity as an essentially rural category, (2) political and economic development priorities which conflict with indigenous interests and needs, and (3) difficulties in promoting access to universal rights and services while simultaneously aiming to promote collective group rights. However, the chapter also reveals that, in Bolivia's and Ecuador's complex and often contradictory environments, some government authorities are already laying the groundwork for more inclusive and pro-indigenous approaches to urban policy and planning.

Following the logic of a variation-finding comparison, the first section of this chapter explores variations in the translation of constitutional content on urban indigeneity and the indigenous right to the city in La Paz. The second section focusses on Quito, while the concluding section compares findings from both cities.

La Paz, Bolivia

As outlined in Chapter 3, Bolivia under the government of President Evo Morales experienced a return of the state. For this reason, Bolivia's national government as well as the local governments of La Paz and Palca, which both compete over political control in the selected case-study sites of Chasquipampa and Ovejuyo, are considered in this chapter as playing a key role in urban governance and, hence, in the translation of constitutional contents on the indigenous right to the city.

Bolivia's national government

Bolivia's new constitution of 2009 allowed a president to serve no more than two consecutive terms in office. According to Bolivia's constitutional court, Evo Morales's first term in office, which began in January 2006, did not count as it had begun before the ratification of the new constitution and the president had not completed a full five-year term. Following ratification of the constitution in 2009, Evo Morales was re-elected as president in December 2009, gaining 64 per cent of the votes. In October 2014, Evo Morales was re-elected with 61 per cent of the votes. In February 2016, a referendum was held asking Bolivian citizens whether the constitution should be amended in such a way as to end existing term limits, thereby allowing Evo Morales to run for a fourth presidential term in 2019. Although the proposed constitutional amendment was narrowly rejected, in December 2017 Bolivia's constitutional court issued a ruling scrapping term limits altogether. At the time of writing, therefore, Evo Morales was expected to run again for the presidency in 2019. This represents a general trend in Latin America whereby incumbent presidents – those of Ecuador, Honduras, Venezuela, and Nicaragua – seek to amend or act against their constitutions in order to eliminate term limits and consolidate power (Landau 2017).

Contradictions and tensions between constitutional rhetoric and actual practice can be observed not only in regard to term limits but also in relation to the respect of indigenous rights. In principle, Bolivia's constitution recognises that all sector policy and planning interventions should address the specific interests and demands of indigenous peoples (see Chapter 2). Bolivia's deputy minister for decolonial affairs confirmed this trend:

> The historical subject of the Bolivian state is no longer the colonial subject or the working class. It is the indigenous person, indigenous like the president, indigenous like the majority of Bolivia's rural and urban population, indigenous as a planetary paradigm. These are the people for whom we [the national government] are working.
>
> (Interview, 21 January 2013)

Indeed, at the international stage, whether that be through speeches at the United Nations (UN) General Assembly or through the organisation of alternative summits on climate change, President Morales promotes indigeneity and principles such as *Vivir Bien* as decolonial alternatives to global development (Canessa 2007). Domestically in Bolivia, government officials make use of indigenous symbols during public events and speeches, whether this be through using the Aymara language, referring to their indigenous nation of origin, or by choice of dress (Postero 2017). Since the rise of the MAS, the national government is increasingly composed of staff of indigenous descent. This includes President Evo Morales himself, who self-identifies as Aymara. The national government also puts emphasis on reducing poverty and breaking socio-economic inequalities and patterns of exclusion which have historically put indigenous peoples at the bottom of Bolivian society. This is mainly achieved through nationalising and expanding key industries such as the hydrocarbon sector. State revenues generated through such activities are used to fund universal social protection schemes, including the pension scheme *Renta Dignidad* and the *Bono Juancito Pinto*, an annual cash grant offered to children to facilitate school attendance by offsetting transportation, food, and material costs (Stefanoni 2012). And indeed, socio-economic reforms have helped rural and urban indigenous peoples and other previously marginalised groups to break out of poverty and to access education as well as jobs in the public and private sectors (McNeish 2013; Postero 2017).

But not all political changes have been positive, especially for specific indigenous groups. In rural Bolivia, for example, the expansion of resource-extraction-related activities often takes place on protected indigenous territories where government authorities either ignore or sometimes violate indigenous peoples' rights to prior consultation (Crabtree and Chaplin 2013; McNeish 2013). Such trends have been noted, for example, in the road-construction project in the Isiboro Secure National Park and Indigenous Territory (TIPNIS) and in current efforts to construct a mega-dam in the Chepete and El Bala regions (Achtenberg 2017; McNeish 2013; Poster 2017).

More importantly, though, constitutional reforms which recognise indigenous rights to the city have not resolved deeply entrenched understandings of indigeneity as essentially a rural category. Instead, historical continuities – or what Bourdieu (1977: 54) calls habitus (practices that are 'a product of history' and operate 'in accordance with schemes generated by history') – explain why urban indigeneity is hardly addressed within national legislation in Bolivia. Interviews and personal communication with government officials, ranging from senior ministers to more junior officials, confirmed that people responsible for the translation of the constitutional rights of indigenous peoples remain guided by a colonial habitus and, therefore, reproduce patterns of coloniality. A deputy minister in Bolivia's Ministry of Culture explained this trend as follows:

I know that we introduce problems with our new laws and reality. Of course, there are many indigenous people living in cities. I live in the city and I would identify as indigenous. But our laws are still drafted by people who follow historical imaginaries and discourses. These people might also be indigenous, but in their minds they are still living in 1952 and want to encourage our indigenous peoples to stay in or return to rural areas where they can work their land. In the city, they believe that people are no longer indigenous.

(Interview, 9 January 2013)

Indeed, history, and not necessarily the new content of the constitution, seems to guide the work of government officials responsible for the draft of new legislation. For example, a deputy minister in Bolivia's Ministry of Autonomies, who played a key role in the formulation of the Law of Autonomies and Decentralisation (LAD) which was ratified in 2010, stated the following:

Indigenous peoples can maintain all their organisational and governance criteria according to their traditions and ancestral knowledge in their own territories. By contrast, in territories in which modernity has been developed—I'm speaking about urban municipalities, the big cities of our department—here we need to have a different focus. The respect of private property and of individual rights according to the liberal model is what we have to stimulate in urban areas. By contrast, in the rural areas and particularly in our indigenous territories, where we as well have individualism but to a lesser degree, we subordinate individual to collective indigenous rights.

(Interview, 16 January 2013)

Similar sentiments were shared by another senior official based in the Ministry of Autonomies (Interview, 16 January 2013): 'Of course, indigenous peoples live in the city, but they cannot claim indigenous autonomy in this area. They have to adapt themselves to the rules of the city.' It is therefore not surprising that patterns of ethno-racial differentiation, which follow colonial understandings of cities as 'non-indigenous' and rural areas as 'indigenous spaces,' are directly incorporated into the LAD which establishes new decentralised units of government characterised by distinct competencies (LAD 2010: art. 8):

1   Autonomous indigenous original peasant (IOP) governments should promote their development as nations and peoples as well as the management of their own territories.
2   Departmental autonomies should drive local socio-economic development and productiveness within their jurisdiction.

3   Municipal autonomies should boost economic, human, and urban developments through the provision of public services to their population. They should also assist rural development.

The LAD also emphasises that each local government unit has to draft a charter (*Carta Orgánica*) which outlines its administrative structure, competencies, and local development priorities (LAD 2010: art. 3).

   Other legislation which translates constitutional rights for indigenous peoples also seems to be guided by a rural bias. For example, following constitutional guidelines (CPE Bolivia 2009: art. 192), the Ministry of Justice ratified a new law on jurisdictional demarcation in 2010 (LJD 2010). The LJD (2010: art. 7) recognises Bolivia as society in which multiple legal systems coexist on an equal basis – these include indigenous justice and ordinary justice. The recognition of indigenous justice allows indigenous communities to rely on their own authorities and legal principles which emphasise collective over individual rights (LJD 2010: art. 13). Yet, like the recognition of IOP territories in the LAD, this law restricts the application of indigenous justice to rural areas in which IOPs live (art. 8, 11). Cities, meanwhile, remain governed by principles of ordinary justice. A senior official in the Ministry of Justice, who was involved in the formulation of this law, gave the following reasons for the spatial restriction to rural areas:

> In the cities we have ordinary justice but some of them [indigenous people] don't want this. Instead, they want to apply indigenous justice. But it is an urban area, a city, right? You can be Aymara or Quechua in the city but you are living in a modern legal environment. By contrast, in the countryside this is different. There you have a communitarian authority and there you can rely on indigenous justice. In the city this is impossible because here we are governed by different rights.
>
> (Interview, 15 January 2013)

Other than reproducing ethno-spatial, rural-urban distinctions, new legislation also tends to draw more on previous laws than on Bolivia's new constitution. This is, for example, the case with the Law of Participation and Social Auditing (LPS) that was ratified in 2013. In its definition of civil society organisations, the LPS follows territorial principles established already in the 1994 Law of Popular Participation (LPP). For rural territories the LPS (2013: art. 14) emphasises that indigenous peasant unions and indigenous communitarian organisations should be involved in participatory and social-auditing processes. By contrast, within cities, participation is restricted to neighbourhood associations – so-called *juntas de vecinos* (JJVVs). Previous research on the LPP emphasised that a focus on *vecinos*, a social category historically associated with 'white' and *criollo* urban residents and currently a way to describe urban residents as neighbours, silenced the issue of ethnic diversity within cities (Goldstein 2013). By reproducing such contents, the LPS once again introduces a participatory model which does not explicitly take into account the issue of indigeneity within cities.

Yet urban indigeneity is implicitly addressed in other legislation. This is particularly the case for the Law to Regulate Property Rights over Urban Estates and Housing (LRPRUEH), ratified in 2012, which emphasises the need to provide city dwellers with tenure rights if they can prove that they have lived on the same plot of land for more than five years (LRPRUEH 2012: art. 10). The LRPRUEH defines *Vivir Bien* as access to material goods such as individual land-tenure rights and housing (LRPRUEH 2012: art. 4). It therefore remains guided by Western property models and recognises only individual rights within cities. Despite such trends, however, a deputy minister working for the Ministry of Public Works, Services and Housing, who was involved in the formulation of the LRPRUEH, further elaborated how the national government understands *Vivir Bien* and addresses indigenous peoples within cities:

> The concept of *Vivir Bien* means improving people's lives through practically transforming the negative and critical conditions that characterise our urban areas. Around one million people do not have secure living conditions. Most of them are indigenous migrants. They lack a land title, or a job, and sometimes even a place to live. We have to respond to the interests of these people.
>
> (Interview, 17 January 2013)

Indeed, then, the LRPRUEH seems to address the interests of indigenous migrants who, as outlined in Chapter 5, demand secure tenure rights and basic public services. At the same time, though, the LRPRUEH ignores the interests and demands of indigenous *comuneros* who want to preserve collective territorial rights.

In a context where indigeneity means different things to different people, policymakers and planners still seem to struggle to come up with a single coherent political and legislative agenda on the translation of the indigenous right to the city. However, it is not only government staff who struggle to promote a new rural and urban development model which focusses on indigeneity, decoloniality, and interculturalism. Bolivian society, in general, is not yet prepared for such transformations, and long-term structural reforms are required. According to a senior member of staff in the Ministry of Education, education plays a key role in this process:

> During 187 years of republican domination and previous colonial domination we developed as a capitalist society. This happened at the expense of our communitarian values which have been repressed through racist politics and discrimination. This generated a negative consciousness among us Bolivians, particularly in cities which were and still are

colonial centres. We cannot reverse this consciousness from one day to the next. It will take some years. I have more hope in our children who will receive a different education and are more likely to change our society.

(Interview, 16 November 2012)

To achieve long-term societal change, Bolivia's national government ratified a new Law of Education (LE) in 2010. The LE is different from previous intercultural education models introduced in the 1990s which predominantly targeted rural indigenous areas. According to the LE, each student, regardless of where they live, should learn about traditional indigenous cosmovision and about modern 'Western' science (LE: art. 3.10). Besides Spanish, each student has to learn one indigenous language in school, ideally the one that is most used in the local area (LE: art. 7). The LE therefore introduces an intercultural (learn about other people's culture), intracultural (learn about one's own culture), and plurilingual (learn Spanish and an indigenous language) education system (LE: art. 3). At first sight, the LE offers a more universal approach to intercultural education, but, during interviews and focus groups, indigenous residents from Chasquipampa and Ovejuyo highlighted that school teachers and local authorities – people to whom Lipsky (1980) would refer as 'street level bureaucrats' – still refuse to address them in their language or to cover their history and traditions in the school curriculum. A gap between national government rhetoric and local practices, hence, seems to remain prevalent. Let us reflect in further detail on how indigeneity is addressed by government authorities at the local level.

Municipal government of La Paz

Since 2010 the political party *Movimiento Sin Miedo* (MSM), in opposition to Evo Morales's MAS, has held a majority in the city council. The municipal government complies with administrative and political guidelines outlined in the 2009 constitution and with associated legislation ratified by the national government. Following the guidelines of the LAD (2010), it ratified a *Carta Orgánica* which addresses constitutional content on indigeneity, development, and interculturalism. Article 1 of the *Carta Orgánica* recognises the intercultural and plurinational character of La Paz. Article 5 recognises the *Chola Paceña* – the indigenous market woman of La Paz – as the official symbol of the city and considers Spanish and Aymara as official languages (La Paz 2012). Otherwise, though, the role of indigeneity seems to be more of symbolic than of practical relevance. The use of Aymara is symbolic in that it is used in banners displayed across the city. In practical terms, Aymara is certainly widely spoken among the local population (see Chapter 5) but not by local government staff. According to the director of La Paz's intercultural unit, only ten per cent of municipal staff speak Aymara (Personal communication, 15 November 2016).

Although the *Carta Orgánica* recognises indigenous cultural symbols and languages, it does not explicitly incorporate the topic of indigeneity into its urban-development planning agenda. Instead, following legislative guidelines in the LPS, the *Carta Orgánica* refers to all inhabitants of the city of La Paz as *vecinos* (La Paz 2012: art. 8). The *Carta Orgánica* also does not draw on the concept of *Vivir Bien* in its definition of development but deploys a human and sustainable development approach which is defined as 'a political, philosophical and institutional concept in which people represent the centre and end of the actions of the autonomous municipal government of La Paz which promises to improve and dignify people's living conditions' (La Paz 2012: art. 8.2).

Local authorities generally offered two reasons why La Paz's municipal government does not rely on *Vivir Bien*. First, mobilising around sustainable development serves the municipality by developing a counter position to the development model promoted by the opposition party – the MAS. Second, and more importantly, municipal staff draw on insights from human and sustainable development because they previously received training on these approaches rather than on *Vivir Bien*. Most senior staff approached in this research mentioned that they received postgraduate training either in the United States or at the Catholic University of La Paz which, since the 1990s, offered public policy and administration courses which followed a so-called 'Harvard Programme' (Kohl and Farthing 2006). The curricula of these courses were strongly influenced by neo-liberal theories and Western planning principles which focus on sustainability, the protection of individual liberties, infrastructure provisioning, wealth creation, and growth. And it is precisely these principles, rather than topics such as indigeneity or alternative development models such as Vivir Bien, which play a central role in public policy and planning practice in La Paz.

A review of La Paz's development planning agenda reveals that the municipal government, between 2010 and 2016, made no explicit mention of the issue of indigeneity. The director of La Paz's human development unit explained this as follows:

> With our interventions we want to improve the quality of life for all residents. As part of our work we do not have a specific focus on indigenous residents but we address them anyway. For example, whether it is a black child or an indigenous child, each gets free meals at school. They can both enter improved healthcare centres. No matter where or for whom, we focus on promoting equality.
>
> (Interview, 7 December 2012)

According to this testimony, the municipal government, like sector ministries that focussed on urban development at the level of national government, followed a universal targeting approach through which it expected to address indigenous residents. Indeed, without explicitly promoting an indigenous rights-based agenda, La Paz's municipal government addressed the demands of indigenous residents for better infrastructure and services (see Chapter 5). In 2012 alone, the municipal government renovated 40 schools – one of them in Chasquipampa – and built eight new healthcare centres and three market halls in peripheral neighbourhoods where the city's indigenous majority lives.

The municipal government also undertakes annual participatory budgeting in every neighbourhood of the city with the aim of implementing small-scale infrastructure projects which reflect local needs. Following the 2013 LPS, the municipality involves JJVVs, and particularly the leaders of these organisations, in participatory budgeting schemes. A civil servant in the district municipality South, where Chasquipampa and Ovejuyo are situated, was certainly aware that a focus on JJVVs ignores specific indigenous organisations. Nevertheless, he argued the following:

> In all of our work, apart the work of the Department of Culture, we do not focus on the topic of ethnicity and the specific demands of groups with different cultural backgrounds. No need to play minority politics. Our infrastructure is for all *vecinos* of La Paz. But don't get me wrong here. Our *vecinos* come from all backgrounds. They include indigenous people who can equally participate in the preparation of the annual operational plan.
>
> (Interview, 27 November 2012)

La Paz's municipal government also prioritises individual over collective rights in planning interventions. This is particularly evident in spatial planning and land management – which La Paz's *carta orgánica* flags as a development planning priority (La Paz 2012). Since 2010, the municipal government has put consolidated efforts into registering land in informal settlements situated in La Paz's peri-urban fringe, including in Chasquipampa and Ovejuyo. Land registration and the provision of tenure status are seen as preconditions for providing services such as water and electricity as well as access to public infrastructure and roads. When registering plots of land and issuing land-tenure rights, the municipal government follows the LRPRUEH and does not recognise collective rights to tenure, thereby ignoring claims by indigenous *comuneros* (see Chapter 5). Between 2012 and 2016, for example, La Paz's territorial planning unit granted tenure rights to 55 per cent of households in Ovejuyo and 80 per cent of households in Chasquipampa (Personal communication with a senior civil servant, 12 December 2016). While the provision of tenure rights was generally viewed as a necessary step to address the basic needs and interests of indigenous residents living in these neighbourhoods, a civil servant in La Paz's territorial planning unit also mentioned another reason for undertaking such interventions:

> Neighbourhoods like Chasquipampa and Ovejuyo are affected by urban expansion. The new people that settled there have very different

demands from the original indigenous owners. Only our municipality can address their interests and not Palca. By providing them with tenure rights these neighbourhoods become part of our jurisdiction and eventually this will allow us to govern in this neighbourhood.

(Interview, 10 December 2012)

In other words, then, granting tenure rights or preliminary residency allowances in areas affected by boundary conflicts with municipal governments such as Palca helps the municipal government of La Paz to gain political control over disputed territories.

Despite arguing that indigenous peoples are addressed through a universal targeting and individual rights-based approach, most municipal government authorities admitted that they were failing to address specific collective indigenous rights mentioned in the 2009 constitution. The director of the municipal development planning unit explained this as follows:

Indigenous rights and interculturalism are topics we have not developed much in our municipality. We really have a limited understanding of what this means in a city like La Paz. The national government talks a lot about the importance of being indigenous or not. Here, we address everyone as *vecinos* but we still have to learn a lot on what to do specifically for indigenous residents.

(Interview, 16 November 2012)

To address some of these knowledge gaps, La Paz's municipal government set up an intercultural coordination unit in 2010. This unit receives resources for four members of staff who, with additional support from Oxfam, co-produced an alternative city plan with indigenous civil society groups from across the city (Sousz *et al.* 2010). This alternative city plan offers guidelines on how to implement specific indigenous rights in an urban context, including the right to collective ownership of rural territories affected by urbanisation, to prior consultation about interventions on indigenous territories, and to culturally appropriate healthcare. The intercultural unit had already started implementing this plan. A new centre for traditional indigenous medicine which provides free healthcare for every urban resident was established in the city centre in 2010. Dialogues were also commenced between the intercultural unit and the municipal authorities to ensure that elements of the plan are mainstreamed into the work of relevant policy sector units. The director of the intercultural unit described these dialogues as the beginning of a long-term struggle:

We can write and discuss a lot with different people here but this will not automatically produce much change. The enemy is in our house.

This is the big problem. We cannot sensitise people or cause change within the municipality if they [municipal staff/planners/policy-makers etc.] claim to know the truth and do not want to listen. They will continue managing urban territories according to their truth. For them, there is no alternative.

(Interview, 10 October 2012)

What the director referred to as 'truth' was outlined in detail previously: It refers to undertaking municipal policy and planning interventions which address the universal interests and rights of all residents including indigenous peoples. As part of such interventions, specific indigenous rights to indigenous justice or collective landownership did not play a central role. Yet such rights are recognised by another municipal government – Palca – which also exercises influence over disputed peri-urban neighbourhoods such as Chasquipampa and Ovejuyo.

Municipal government of Palca

Palca claims most parts of Ovejuyo and Chasquipampa as its jurisdiction, thereby making a significant claim for political and territorial control over peri-urban La Paz (Personal communication with Palca's mayor, 12 November 2016). In addition to these disputed peri-urban areas which are home to around 82,000 residents, the municipality of Palca contains 16,000 inhabitants who are predominantly rural peasants of Aymara origin (INE 2014). The municipal council is controlled by a majority of members of the MAS, the same political party which is in control of Bolivia's national government. Palca's municipal government employs 35 members of staff and relies on an annual budget of approximately 17 million Bolivianos. The absence of financial and human resources has direct repercussions on Palca's ability to provide basic public services to its citizens. According to a World Vision (2012) study, 98.7 per cent of Palca's rural residents lack access to basic physical and socio-economic infrastructure, such as paved roads, water, or sanitation.

Unlike La Paz, Palca's municipal government does not have a municipal development plan or *carta orgánica* outlining its policy and planning agenda. Instead, according to personal communications with municipal staff, Palca draws on the new constitution and relevant national legislation to define and justify its interventions. Within its jurisdiction the municipality recognises collective land-tenure rights and specific indigenous rights such as indigenous justice. The mayor of Palca justifies this as follows:

Brother Evo [referring to President Evo Morales] and his government clearly say in the constitution that our indigenous original peasants can govern themselves according to their own principles and law. This is what we allow our residents to do in our municipality. What else should we do? The few people that work in the municipality do not have the resources to exercise control.

(Interview, 18 January 2013)

As is evident in this testimony, the municipal government not only relies on an indigenous rights-based agenda because of the presence of a predominantly indigenous population. It also applies such an agenda for more pragmatic reasons. By recognising specific indigenous rights to self-governance, the municipal government can delegate responsibilities, such as the provision of public services, directly to its residents. Such tendencies were confirmed in an interview with the director of the social development unit of the departmental government of La Paz (Interview, 21 January 2013): 'Rural areas like Palca do not have the capacity to govern their people. We observe that in these areas governments simply allow their communities to govern themselves.'

The municipality does, however, exercise influence not only over its rural territories but also over peri-urban areas like Chasquipampa and Ovejuyo. Reasons for maintaining or regaining control over these areas were outlined by Palca's mayor himself, an indigenous resident living in these neighbourhoods:

We are all indigenous peoples here. We used to live and still live in a community, like our ancestors. Throughout the last decades they [land invaders, new residents, the municipality of La Paz, etc.] tried to break our traditions and brought in individualism. They discriminated and marginalised the ancestral residents who used to cultivate their land here. This is our land and we govern it according to our principles.

(Interview, 18 January 2013)

Adding to the mayor's account, another civil servant working in the municipality offered additional motivations for Palca's territorial claims in peri-urban territories (Interview, 17 January 2013): 'We lack resources because La Paz stole our land. We are about to get back what belongs to us. By regaining political control over this area we can increase our budget.' Instead of reinventing indigenous traditions, a core interest for Palca's municipality, according to this testimony, is to increase its income – something which is achieved by collecting property taxes from residents living in more densely populated urban areas.

In order to substantiate its political and administrative influence, the municipality of Palca constructed its town hall in Ovejuyo. From here, Palca grants land-tenure rights and permissions for land subdivision as well as construction permits not only to *comuneros* but also to newly arriving residents or private enterprises interested in investing in the area in exchange for service charges. In contrast to La Paz, though, the municipality of Palca also legally recognises collective tenure rights within peri-urban areas, particularly for *comuneros* who used to inhabit these lands prior to urbani-sation. A civil servant explained this as follows (Interview, 17 January 2013): 'Of course we respect collective land ownership in Chasquipampa and Ove-juyo. It is part of our rural jurisdiction and we apply the same rights there as elsewhere.' Like civil servants in the municipal government of La Paz, staff in Palca's municipality refer to existing legislation on land tenure to justify their political agenda on land management and administration. But, in con-trast to La Paz which follows the LRPRUEH and mainly respects individ-ual tenure rights, Palca draws on the contents of the new constitution and on earlier legislation such as the 1953 Agrarian Reform decree. Following the 1953 reforms, ex-*hacienda* lands were redistributed to peasant families. Each family thereby received three acres of land and together they jointly owned five acres of communal land. As highlighted in interviews with dif-ferent members of staff, the municipal government of Palca recognises these tenure rights up until now but, in exchange for a service charge, also grants permissions for subdivisions if occupiers intend to sell parts of their land.

In summary, then, this section has demonstrated that with some notable exceptions such as La Paz's intercultural unit, most government authori-ties do not explicitly focus on urban indigeneity in policy and planning in-terventions. Rather, they remain guided by understandings of the city as non-indigenous and modern place, follow constitutional and legislative guidelines which replicate rural understandings of indigeneity, seek to gain political control over previously semi-autonomous indigenous territories af-fected by urban expansion, or struggle to respect collective indigenous rights while simultaneously ensuring that urban residents – including indigenous peoples – can access universal rights and services. All this demonstrates that there exists a variety of approaches towards addressing the indigenous right to the city, with some directly responding and others ignoring the in-terests and needs of urban indigenous residents. Let us now explore how the indigenous right to the city is addressed by government authorities in Quito.

Quito, Ecuador

As in Bolivia, it is possible to detect a return of the state in Ecuador and, in Quito, the national government as well as the city's municipal government plays a key role in urban governance and, hence, in translating constitu-tional rights such as the indigenous right to the city.

National government

Following the ratification of Ecuador's new constitution in 2008, Rafael Correa's party PAIS Alliance (AP) was re-elected with 52 per cent of the votes and won 69 out of 124 seats in Ecuador's parliament. Correa was re-elected for a second presidential term in 2013. In 2017, the AP candidate,

Lenin Moreno, formerly Correa's Vice President, was elected President. As in Bolivia, therefore, Ecuador's AP has been in control of the national government since the ratification of the new constitution. The national government translated constitutional contents on indigeneity into national development plans and legislation but, as will be demonstrated later, this has barely impacted policy and planning practice as the government has followed other political priorities.

Both the 2009 and the 2013 national development plans highlight the importance of respecting and strengthening group rights, including those of indigenous peoples in rural and urban areas (SENPLADES 2009, 2013). Senior members of staff in Ecuador's National Secretariat for Development Planning (SENPLADES) further confirmed in a personal written communication (26 April 2016) that *Buen Vivir* relates to the 'recognition of vulnerable groups—including women, indigenous peoples, and people with disabilities—and the right to the city.' In order to effectively translate new constitutional rights within Ecuador's diverse regions, cities, and villages, the national government deepened decentralisation. The new decentralised structure of the state is outlined in the new Organic Code of Territorial Organisation, Autonomy and Decentralisation (COOTAD) which was ratified in October 2010. The COOTAD (2010) gives Ecuador's national government the responsibility to draft new legislation which sets guidelines on how specific constitutional content, such as intercultural education (CPE Ecuador 2008: art. 343), intercultural healthcare (art. 358), or intercultural urban development (art. 375), should be addressed within all cities. Further, to ensure that sector ministries take into account the topic of indigeneity, Ecuador's government drafted a new Law on Citizen Participation (LCP) in 2010 which guarantees to involve indigenous peoples and their collective organisations and respective civil society and state institutions within processes of participation (art. 1) and social auditing in all sector policies (art. 46). The LCP (2010: art. 81) also recognises the right to prior and informed consultation on interventions taking place in indigenous territories.

Contrary to the LCP (2010), however, the national government limited the influence of existing institutions that represent the specific concerns of indigenous peoples. A common explanation for these gaps between legal rhetoric and practice is that Ecuador's national government, in reality, follows distinct political priorities. A senior official from SENPLADES explained the rationale that underpins gaps as follows:

> Our government mainly works for Ecuadorians citizens. The indians are a minority. As in any democracy, a minority does not rule. We treat indians as equals with the same universal rights and services. Unlike previous governments, we no longer want to have this politics of difference. Our history is a history of difference. Now we want to be one

people governed by a strong government. For this reason, we decided to close those institutions that were controlled by opposition forces and focused only on minority groups.

(Interview, 10 June 2013)

In practice, Ecuador promotes a political agenda which prioritises the universal rights of all citizens – including indigenous peoples – over specific group rights. As in Bolivia, Ecuador's national government decided to enhance national control over the economy and to expand resource extraction with the aim of generating additional resources which can be channelled into the provision of universal basic services and social protection schemes for the elderly, people with disabilities, or children (Horn and Grugel 2018). The overall aim of such interventions, hence, is to reduce poverty and inequality. And indeed, between 2006 and 2016, poverty levels fell from 36.5 per cent to 22.5 per cent and inequality decreased from a GINI score of 0.55 to 0.47 (Weisbrot *et al.* 2017).

As part of its socio-economic policy agenda which focusses on generating state income by expanding the resource-extraction sector, the government decided to sell mining rights over land held by indigenous communities to foreign companies, bypassing internationally agreed mechanisms on community rights to prior consultation, as well as closing down institutions and programmes targeting affected indigenous peoples. For example, in 2013, the national government closed the Coordinating Ministry of Patrimony (CMP) and transferred its staff to the Ministry of Culture. In addition to the CMP, the government also closed the Directorate for Intercultural and Bilingual Education (DINEIB) and the Secretariat of Peoples. The latter was integrated into a new National Secretariat of Policy Management which coordinates the implementation of constitutional principles such as *Buen Vivir* for all sector ministries. A senior civil servant who used to work in the Secretariat of Peoples highlighted some of the consequences of the closure of the institution:

> Less than 25 percent of our staff relocated to the new ministry and out of those even fewer continue working on indigenous questions. In the new ministry, the government wants to monitor how *Buen Vivir* is achieved for all citizens. Unlike in our Secretariat of Peoples, indigenous affairs play only a minor role on their agenda.

(Interview, 16 May 2013)

A reduction in the influence of institutions that address indigenous peoples could also be observed at the Council for the Development of Indigenous Nations and Peoples of Ecuador (CODENPE). Established prior to Correa's government in 1998, CODENPE was tasked with monitoring indigenous affairs and implementing specific indigenous development programmes, such as promoting microcredits and saving schemes within indigenous communities. In the financial year 2012, CODENPE was still allocated the equivalent of USD 12 million of government funding to implement such interventions. Yet, according to information provided by the director of CODENPE (Interview, 9 April 2013), Ecuador's national government cut the budget of this institution by more than two-thirds in 2013. In 2016, CODENPE was scheduled for closure (Horn and Grugel 2018).

While institutions responsible for assisting and monitoring the implementation of indigenous rights across policy sectors were closed, other sector ministries still followed the constitutional mandate and addressed indigenous rights and intercultural principles in their work. This is particularly the case with education and healthcare but less so with urban development. With regard to education, the national government ratified a new Law on Intercultural Education (LIE) in 2011. Article 2z of the LIE states that interculturalism refers to 'unity in diversity. It strengthens intercultural and intracultural dialogue and valorises the practices of different cultures which stand in harmony with human rights.' The LIE (2011: art. 5) highlights that intercultural education is applicable to all Ecuadorian citizens. Therefore, the Ministry of Education is obligated to develop a new national curriculum that incorporates ancestral languages and indigenous history. In addition, depending on the regional context, local schools have to incorporate specific cultural elements that characterise a particular region, village, or city (LIE 2011: art. 6). As part of these educational reforms, the national government opened a new unit for intercultural education within the Ministry of Education. Schools which were part of the bilingual education system introduced by previous governments were reintegrated into the new national intercultural education system as long as their staff could prove or obtain relevant qualifications to work as teachers. Bilingual schools which could not comply with this criterion were allowed to continue operating only when situated in remote rural areas or marginalised urban neighbourhoods in which, at the time of the ratification of the new law, no other educational centres operated. In the context of Quito, this meant that bilingual schools such as the Chaquiñán College, which operated on a voluntary basis, had to close (see also Chapter 7).

The attempt to incorporate indigenous languages and intercultural topics within universal and state-run education schemes led to new practical challenges. A senior civil servant in the Ministry of Education explained this as follows:

> Society is not yet prepared and neither are the people to decide who should make these changes. In our ministry people responsible for writing new curricula don't know much about indigenous history or languages. In our schools, particularly in cities like Quito, teachers and students are often racist. They don't speak or want to speak indigenous

languages. They believe indigenous peoples are not part of the city. We have started training the teachers but it will take years before our reforms have an impact on the ground.

(Interview, 7 May 2013)

Ongoing problems of racism and discrimination within cities have been reported not only by most indigenous research participants (see Chapter 5) but also by government staff that worked on addressing urban indigenous peoples and intercultural principles within other policy sectors. For example, similar to reforms emanating from the Ministry of Education, Ecuador's Ministry of Health followed the constitutional mandate to introduce intercultural healthcare services (CPE Ecuador 2008: 358). For this purpose, the ministry opened a new intercultural health unit which was declared responsible for monitoring whether hospitals and healthcare centres employ staff that speak indigenous languages. Particularly within cities, little progress has been made on this topic. According to the director of this unit (Personal communication, 24 April 2016), by 2016, less than ten per cent of city hospitals hired staff who spoke indigenous languages. The director also pointed out that many doctors and nurses remain guided by racist attitudes and are unwilling to attend language courses offered by the Ministry of Health (Interview, 6 June 2013): 'Most of the doctors tell us that before they learn an indigenous language, the indigenous peoples should themselves start to learn the language which is spoken in the city – Spanish.'

Ecuador's Ministry of Urban Development and Housing (MIDUVI) particularly struggles to incorporate topics such as indigeneity. Following Ecuador's new constitution, MIDUVI defined *Buen Vivir* through the notion of the right to the city, acknowledging specific indigenous rights to the city (see Chapter 2). Yet, to date, MIDUVI has not defined how specific indigenous rights to the city can be addressed in practice by the government. A senior civil servant in this ministry explained the matter as follows:

We realise that cities are the places which are the most diverse. Most of the Ecuadorians live there. They come from all parts of the country. Many of them belong to indigenous peoples or nations to whom our constitution grants new rights. These groups were never taken into account by us. The constitution requires us to develop specific programmes that address indigenous peoples and other groups but at this moment we simply lack knowledge on how to do this. Most of us here are civil engineers or architects who were never trained on cultural issues. We simply do not know what interculturalism and indigenous rights mean in relation to our work, which centres on providing housing and urban infrastructure to people living in cities.

(Interview, 5 June 2013)

Instead of focussing on specific indigenous 'rights to the city,' MIDUVI mainly focusses on guaranteeing urban residents their universal right to housing and public infrastructure. Since Rafael Correa became President in 2007, the government significantly increased its annual budget for housing programmes, housing subsidies, and urban public infrastructure investment (Horn and Grugel 2018). According to civil servants in MIDUVI and SENPLADES, indigenous peoples indirectly benefited from such interventions. A senior civil servant in MIDUVI explained this as follows:

> Our policies target the entire population of Ecuador but particularly those who are the most vulnerable. By coincidence, indigenous peoples but also Afro-Ecuadorians belong to this group. They are often in receipt of our housing subsidy schemes or social housing programmes.
>
> (Interview, 5 June 2013)

The fact that urban indigenous peoples indirectly benefited from universal targeting schemes was confirmed in other interviews. A senior civil servant from SENPLADES, for example, stated:

> *Buen Vivir* is a universal idea. If indigenous people in our cities live badly they should have access to universal rights for housing, public services, or education. We don't want them to be poor; we simply want them to live in conditions that equal those of every other citizen.
>
> (Interview, 10 June 2013)

According to statistical evidence provided by SENPLADES, only 47 per cent of urban indigenous peoples had access to adequate housing in 2006. This increased to 84 per cent in 2016, which again confirms the trend that the specific interests and demands of indigenous peoples are mainly addressed through universal policy interventions (Personal written communication with SENPLADES, 26 April 2016).

Municipal government of Quito

Ecuador's COOTAD (2010, art. 4) also specifies the competencies and responsibilities of local government units. The COOTAD significantly broadens the competencies of local governments which previously were mainly responsible for the provision of physical infrastructure. Municipal governments are now also responsible for undertaking spatial planning, boosting local economic development, protecting intercultural diversity, planning and regulating transport, and improving local healthcare and security. In accordance with the constitution and new national legislation, municipal governments must also ensure that diverse residents – including indigenous peoples – are involved in the preparation, implementation, and evaluation of public policies and urban planning interventions (COOTAD 2010: art. 84). In addition, Article 93 of the COOTAD grants indigenous territories and communes administrative competencies to implement indigenous rights which are outlined in Article 57 of the new constitution (CPE Ecuador 2008). Unlike previous legislation which defined communes as rural territories, the COOTAD (2010: art. 24) also recognises the presence of communes within urban areas: 'Where communes exist whose total or partial territory is situated within urban areas, cantons or autonomous municipal districts must, via their legislative bodies, consult and involve these communes in the governance of their community land and goods.'

This section discusses to what extent and how local authorities in Quito comply with the COOTAD and address the specific interests and demands of indigenous residents. Similar to La Paz in Bolivia, with more than 2,000 members of staff and an average annual budget of more than USD 500 million, Quito's metropolitan district municipality was by far one of the richest local governments, in terms of financial and human resources, in Ecuador. From 2009 until February 2014, Rafael Correa's AP party held a majority in Quito's municipal council. In February 2014 the AP, led by Mayor Augusto Barrera, lost the municipal election to the opposition candidate Mauricio Rodas who represents the party '*Movimiento Sociedad Unida Mas Accion*' (SUMA). The Barrera and Rodas administrations adapted slightly different approaches towards urban indigeneity.

Under the legacy of Augusto Barrera, Quito's municipal government followed a similar approach to Ecuador's national government and prioritised universal over specific indigenous rights. For example, the director of Quito's municipal housing enterprise insisted that 'There is no differentiation in our social housing approach for indigenous people. We treat housing as a universal human right and run housing projects for indigenous families, *mestizo* families, and everyone else' (Interview, 29 May 2013). Indeed, between 2012 and 2013 alone, the municipal government implemented social housing projects in predominantly indigenous peripheral neighbourhoods and provided approximately 10,000 residents with a new home (DMQ 2012c).

During Barrera's period as mayor, Quito's municipal government ignored specific indigenous rights when they conflicted with the government's economic development agenda. This trend was visible in the city's airport project as well as in attempts to revitalise parts of the city centre. With national government support, the municipal authorities completed the construction of Quito's new airport in 2012. To attract international businesses and to generate employment opportunities, the municipal authorities also planned the construction of three new industrial parks to be located directly next to the airport. However, these large-scale infrastructure projects took place on the territory of a number of suburban indigenous communes who – according to the new constitution and the COOTAD – had the right to be consulted about interventions taking place on their territories. According to political leaders of these communes, the municipal government failed to comply with this legislation. This was acknowledged by the municipal authorities themselves. A civil servant in Quito's territorial planning unit, for example, stated that, 'With the new airport project we generate jobs and improve the lives of all residents including indigenous ones. The improvement of lives comes with a cost. You cannot address everyone as equal all the time' (Interview, 22 May 2013). As a consequence, citizen involvement on topics like the airport project remained selective, with invited spaces – to use Miraftab's (2009) terminology – being closed off to those people who inhabited land dedicated for redevelopment. The director of Quito's participation unit put it this way (Interview, 21 May 2013): 'As a municipality we would not involve people who are against our agenda. Why would we involve *comuneros* in the airport project? To make life easier, we involve only those people who support this project in the first place.'

Similar tendencies could also be observed in the city's central San Roque market, a hub for indigenous migrants like those assembled in the AECT-Q (see Chapters 5 and 7). In co-operation with the Ministry of Urban Development and Housing, Quito's municipal government sought to revitalise this part of the city and to make it attractive for private investors and tourists. As part of this revitalisation effort, the municipal government intended to close the central San Roque market and relocate indigenous vendors to other parts of town. The justification for replacement was provided by a member of staff in Quito's commercial unit (Interview, 10 July 2013):

> The area has a lot of potential for tourism and private investment. The indigenous people disturb this development. In this city no one should disturb anyone else. My right to the city stops once I violate the right to the city of others.

Even though Barrera's municipal government ignored specific indigenous rights, interests, and needs in its economic development interventions, it addressed them in healthcare and cultural interventions taking place in neighbourhoods with an indigenous majority. Unlike in La Paz, where the authorities sought to mainstream intercultural affairs into the work of all local government units, Quito relied on a policy-targeting approach. An example of such a targeted intervention was the healthcare programme '60 y Piquito.' As part of this programme, the municipality of Quito provided workshops and courses on healthcare for people aged over 60. These workshops were held in more than 120 local community centres across the city. Zonal administrations were responsible for implementing the programme in such a way that the specific cultural demands of residents were taken into account. The director of the zonal administration of Calderon described what this meant in practice:

> In the communes but also in some neighbourhoods most of the elderly attending '60 y Piquito' are indigenous. To communicate with these people, we hire staff who speak Kichwa or we work with community residents who can translate to the elderly what our community workers are saying to them.
>
> (Interview, 29 May 2013)

Quito's cultural sector unit relied on a similar approach. It requested zonal administrations to identify the particular cultural characteristics of each neighbourhood and to fund events that responded to resident's interests. In neighbourhoods predominantly inhabited by indigenous inhabitants, the municipality funded traditional cultural events such as 'Inti Raymi' – the festival of the sun which is celebrated annually in June. According to information provided by a member of staff in Quito's Secretariat of Culture, the municipality allocated the equivalent of over USD 160,000 to indigenous community organisations in more than 30 neighbourhoods to enable these organisations to run folkloric festivals themselves and according to their specific interests. This certainly helped indigenous residents to revitalise their ancestral traditions in the city.

Unlike Barrera, the administration of Mayor Mauricio Rodas took a slightly more pro-indigenous approach. A member of Quito's development planning unit explained that this shift could mainly be explained as a reaction against pressure from different indigenous residents:

> For the national government, indigenous peoples represent 4 percent of the country. In Quito, they represent only 7 percent but that number is different for some districts. Especially in the Northern and Southern periphery, they sometimes represent as much as 60, 70 or 80 percent of the local population. They increasingly organise around their indigenous background and mobilise in support of their constitutional rights. The leaders of indigenous organisations have come to the municipal offices again and again, over the years, to present specific demands. After a while the municipal government simply had to respond. Market vendors pushed us to shift our approach towards San Roque. Following many complaints, we also decided to work more closely with the members of the different communes.
>
> (Interview, 20 October 2016)

In San Roque the municipal government departed from a focus on relocation and, instead, started coordinating with indigenous merchants to undertake an incremental modernisation of the market hall, including the installation of new toilets and doors, as well as restoring the walls and floors and repairing the sewage system. These changes followed long consultations and better addressed the interests of market vendors. This was highlighted by a leader of the AECT-Q, who stated:

> Finally they have started to respect our right to prior consultation. The municipality no longer tells us what will be done but asks us what we want them to do. We are grateful that they listen and hope that they will no longer plan to displace us.
>
> (Interview, 20 April 2016)

In 2016, the territorial planning unit of Quito's municipal government also decided to hire a dedicated member of staff whose task it is to identify practical ideas for integrating collective indigenous rights to land in communes within municipal land-use plans which have traditionally recognised only individual tenure. Such an approach is supposed to better comply with the COOTAD law which requires municipal governments to consult and involve communes in planning processes affecting their territories. A first initiative by this member of staff was to establish a co-government unit which liaises with representatives from Quito's 49 communes over land-use registration. This works as follows: In the initial planning stage, leaders of a *cabildo* must approach every household in their commune to register plots of land and tenure type, taking note of both individually and collectively owned plots of land as well as tenants and structure owners. Following this community mapping exercise, plots should be mapped out virtually through geographic information software (GIS) (Personal communication with the director of Quito's co-governance unit, October 2016). In the event that two different plots overlap, leaders of the *cabildo* and the municipal co-governance unit should engage in dialogue with the affected parties. Upon completion of a land-use map for a commune, the land-use registration process moves on to the consultation stage. In order to be registered by the municipality of Quito, land-use registration maps require approval from every commune resident over the age of 18 as well as from the leadership of the *cabildo* and the director of Quito's co-governance unit (Personal communication with the director of Quito's co-governance unit, October 2016). The land registration process was started in April 2016 and, to date, no plan has been finalised. Nevertheless, residents of communes perceive the process as positive. This is evident, for example, in a testimony given in 2016 by Enrique, an elderly migrant from Llano Grande who, when initially approached in 2012, had been sceptical towards the municipal government:

> Finally, they are taking us seriously in Quito. Of course, we still have problems and some people in our commune want different things. But the registration process makes us talk to each other. I am learning a lot from my neighbours and there is at least one benefit. We have become humbler and more accepting of our differences. Maybe that is

something we had previously forgotten. Everyone wanted something different. I wanted to preserve my agricultural land. The youngsters wanted to leave it behind. We didn't talk to each other and, for this reason, we did not find consensus. Finding consensus is still difficult, but by talking to one another we will get there. Perhaps we will come together through this process as different people with different interests who nevertheless share common values of being and living in the commune of Llano Grande, each one of us holding onto our specific way of life while respecting those of our neighbours.

(Interview, 20 October 2016)

In summary, then, the aforementioned examples illustrate how Ecuador's national government and Quito's municipal government address indigenous interests and needs in distinct ways. The national government as well as Barrera's municipal government ignores and, at times, violates specific indigenous rights to address other political priorities, namely the promotion of economic development initiatives. Otherwise, urban indigeneity is addressed mainly through a universal rights-based approach which focusses on promoting *Buen Vivir* among all citizens, including indigenous peoples. This does not mean, however, that government interventions are monolithic. National and government municipal authorities have started to address specific indigenous interests and demands in sectors such as culture, healthcare, and social housing. Furthermore, since the election of Rodas in 2014, Quito's municipal government has become even more enthusiastic in addressing the diverse interests and demands that are articulated by different migrant communities as well as by *comuneros*.

Conclusion

The indigenous right to the city as outlined in Bolivia's and Ecuador's constitutions is not always being incorporated into urban policies and planning interventions in La Paz and Quito. This can be explained by the fact that government authorities responsible for implementing the constitutions are themselves social actors whose work is influenced not only by legal documents but also by personal views and the specific institutional and structural environments in which they operate. The findings presented in this chapter confirm this trend.

In La Paz, constitutional content on urban indigeneity and intercultural urban development often conflicts with the personal views of government officials, which remain guided by a colonial habitus and a static and spatially fixed understanding of indigeneity as a rural category associated with tradition and backwardness, and by a perception of the city as a 'white,' modern, and non-indigenous place. However, such a historically rooted explanation does not explain gaps between constitutional rhetoric and urban policy and planning practice in Quito. Instead, here it is rather the fact that constitutional rights on indigeneity conflict with other political priorities of government authorities such as promoting large-scale economic development programmes.

In both cities, it is possible to observe how the authorities struggle to promote access to universal rights and services while simultaneously respecting the interests and collective rights of specific urban indigenous groups. Nevertheless, it is important to emphasise that the application of universal and individual rights-based approaches, and the use of Western planning models often still yield positive results, contributing to poverty-reduction among the urban indigenous population and responding to the interests and demands for individual tenure, housing, and basic public services raised by many urban indigenous residents (see Chapter 5).

This chapter also revealed that different institutions involved in urban governance in La Paz and Quito are composed not of monolithic but of heterogeneous social actors. While a majority of government authorities seem to struggle with the idea of the indigenous right to the city, others – such as La Paz's intercultural unit – seek to incorporate constitutional content on indigeneity into their urban policy and planning practices. By introducing an intercultural education system that is universally applicable, both national governments aim to confront and break deeply entrenched ethno-racial divisions and, instead, promote encounter and interaction between different people of distinct cultural backgrounds. While such reforms are unlikely to have short-term impact, they do represent a first step towards the long-term structural and societal changes required for the making of a more open, tolerant, plurinational, and intercultural society.

This chapter has also revealed how approaches towards the translation of constitutional rights such as indigenous rights to the city not only vary according to context but also change over time. This is the case in Quito, where the current municipal government has adopted a more pro-indigenous approach when engaging with indigenous market vendors in San Roque and *comuneros*. As regards to the latter, the municipal government developed a land registration approach which aims (1) to respect individual and collective rights in the planning process; (2) to combine mechanisms of indigenous governance as promoted by *cabildos* with direct democracy (e.g. through popular consultations) and conventional urban governance as promoted by Quito's municipal government; and (3) to carefully negotiate internal conflicts around land use within communes. Such a co-creative approach towards urban indigenous policy and planning practice not only breaks with previous exclusionary state-led practice but also problematises and productively addresses differences within indigenous communities.

To conclude, then, similar to understandings of indigeneity among self-identifying urban indigenous residents, there does not exist one coherent government response towards the indigenous right to the city. Rapid and scalable solutions on this topic are unlikely. Likewise, what works in one (part of the) city might be difficult to implement in another (part of the) city. For this reason, it is perhaps better to abstain from drawing out general policy recommendations. Instead, there is a need to continue the search for practices that work best in the specific context of individual cities and subsequently to identify pathways on how to strengthen and deepen such interventions. This not only requires paying close attention to the policy and planning practices of national and local governments. As will be demonstrated in the next chapter, much can also be learned from the everyday practices of urban indigenous residents.

7 Claiming indigenous rights to the city

Introduction

The previous chapters established what urban indigenous residents want in La Paz and Quito and to what extent and how national and local government authorities address the interests and demands of these people through policy and planning interventions. This chapter now examines the role of indigenous residents as planners of their own lives who seek to claim their indigenous right to the city through a variety of self-help, negotiation, and contestation practices.

Through case-study illustrations from La Paz and Quito, this chapter finds that it is not every ordinary resident but mainly indigenous leaders, most of them older men, who play a key role in claiming the indigenous right to the city from below. It also argues that indigenous leaders do not always represent the interests of all their community members, especially those of women and young people. Such practices are, however, not uncontested. This is highlighted through the discussion of the work of an indigenous youth tribe in La Paz which seeks to confront uneven power relations and patterns of exclusion within indigenous communities, thereby offering an example of a more inclusive and emancipatory struggle for the indigenous right to the city.

Following once again the logic of a variation-finding comparison, the first section of this chapter discusses how different people involved in a variety of residential organisations in the neighbourhoods of Chasquipampa and Ovejuyo in La Paz claim their indigenous rights to the city. The second section focusses on Quito, paying particular attention to the work of one indigenous community-based organisation (CBO) – the AECT-Q (see Chapter 4). The final section compares the findings from both cities.

Everyday indigenous urbanism in Chasquipampa and Ovejuyo

Most residents in Chasquipampa and Ovejuyo voice their specific interests and rights-based claims through engagement in multiple CBOs, including indigenous peasant unions, neighbourhood associations (JJVVs), folkloric associations, and youth tribes. Unlike in the 1990s and early 2000s, when indigenous CBOs engaged in insurgent urbanism (Becker 2010; Lazar 2008), in today's political context they mainly negotiate with different government authorities and/or engage in participatory processes. For example, in order to protect rural plots of land, residents join indigenous peasant unions that mainly engage with the municipality of Palca. At the same time, though, to improve public infrastructure services and road access to their urban homes, residents would need to be members of a JJVV which would raise these topics with the municipality of La Paz as part of the participatory budgeting process (see Chapter 6). Through such interactions, indigenous residents manage to fulfil their interests and aspirations. To unpack this further, the remainder of this section provides four illustrations that describe the work of different CBOs in Chasquipampa and Ovejuyo.

Neighbourhood associations

Migrants and *comuneros* alike demand individual tenure rights which would allow them to access urban infrastructure and public services provided by the municipality of La Paz (see Chapter 5). The tactics used by residents and, in particular, by community leaders to access urban land, claim tenure rights, and receive public services have changed significantly over time (see Table 7.1). Local residents generally highlighted that, during the initial period of neighbourhood formation in the 1980s, the state was absent from the neighbourhoods. Without interference from national or local government authorities, *comuneros* sold parts of their land to land speculators or directly to newly arriving indigenous migrants. New settlers lacked land titles that demonstrated their landownership. The only document they normally received was a confirmation of payment (*minuta de compra*) from a land speculator. To avoid land invasions by other incoming migrants, new residents quickly built a house or temporary structure on their plot, and subsequently ensured that at least one household member always remained vigilant on the plot.

In the 1980s and 1990s, residents founded their own urban neighbourhood organisations in order to collectively improve life in their new neighbourhoods. The organisational structure of these new neighbourhood organisations often resembled that of traditional rural indigenous communities. Pascual, a migrant and member of a JJVV in Chasquipampa, explained this as follows (Interview, 1 November 2012): 'Everything worked like in the countryside. We rotated our leaders. We held many assemblies to discuss our problems collectively.' Neighbourhood associations, in this period, also relied on traditional indigenous practices such as the *mita* (collective work schemes). Jose – migrant and member of another JJVV in Chasquipampa – described this as follows:

Table 7.1 Accessing tenure rights and public services in Chasquipampa and Ovejuo

Year	Regulation	Implementing government unit and key contents	Tactics used by indigenous residents and leaders
Until 1994	None	None	• Self-help: Ownership through occupation • Rely on indigenous governance principles in neighbourhood organisations
1994	Law of Popular Participation (LPP 1994)	Municipality of La Paz (municipal law 1669 – La Paz 1995) issued individual titles for land occupants	• Adopt the organisational criteria of JJVVs outlined in the LPP • Adopt 'white' *vecino* identity and hide indigenous features • Establish clientelist ties with government representatives
2012	Law to Regulate Property Rights over Urban Estates and Housing (LRPPUEH 2012)	National government and municipality of La Paz provided individual titles to land occupants	• Adopt 'white' *vecino* identity and hide indigenous features • Establish clientelist relations with government unit

Source: Elaborated by the author.

In the 1980s my dad was one of the founders of this neighbourhood organisation. At that time, I was five or six years old. There were no roads to get into the city. So, my dad and other neighbours did *mitas* every weekend. They worked like in my rural community of origin. They used dynamite to pave out the land. All this here was a rocky hill until we made it a road. Thanks be to God that no one died during this work and that there were no severe accidents.

(Interview, 12 November 2012)

While the first neighbourhood organisations relied on traditional practices, such patterns changed in later stages of neighbourhood consolidation. According to Francisco, a *comunero* who is a member of a JJVV in Chasquipampa and also of an indigenous peasant union in Ovejuyo, a crucial turning point in the history of collective organisation in the neighbourhoods was the ratification of the Law of Popular Participation (LPP) in 1994:

To move forward as a neighbourhood organisation, we adapted to what the law 1551 [LPP] says. This law told us that all of us should be *vecinos* and that we have to organise ourselves democratically. There should be only a few leaders per neighbourhood organisation which we elect at the polls every two years. The leaders have to follow the rules. They can hand in suggestions for the annual operational plan. If suggestions are taken up, the municipality of La Paz will do infrastructure works for us.

(Interview, 19 November 2012)

Legislative changes, hence, compelled neighbourhood organisations to change their internal structure. The old neighbourhood organisations became JJVVs, and members had to emphasise their role as *vecinos* and work with the municipality of La Paz, which, from this point onwards sought to exercise political control and promote urban development in the neighbourhoods. As was outlined in Chapter 6, to substantiate its influence, La Paz issues local residents with tenure rights and invites the leaders of JJVVs to take part in annual participatory budgeting processes. JJVV membership is restricted to those residents who obtain an individual land title (Personal communication with a member of staff in La Paz's territorial planning unit). According to the LPP, only one JJVV can represent a specific territory and, until the present, there are three JJVVs which are active within the neighbourhoods. These JJVVs are members of a wider citywide network of JJVVs – the federation of neighbourhood organisations (FEJUVE).

JJVV leaders play a central role in negotiating demands for access to public services and infrastructure works. They generally do this through their formal involvement in annual participatory budgeting procedures organised by the municipality of La Paz. Pascual explained this process as follows (Interview, 1 November 2012):

Since I have been a member of this JJVV, the infrastructure works have always been approved and implemented by the municipality of La Paz. You go to the yearly meeting and they put your demand into the annual operational plan. That's it.

Other than requesting infrastructure works in participatory budgeting exercises, leaders of JJVVs have recently been involved in new participatory spaces which the municipality of La Paz introduced in their new *Carta Orgánica*. These include, among other events, assemblies during which municipal staff present new large-scale urban-development projects to civil society representatives. Assemblies are, in principle, open to the participation of different CBOs but, in practice, only JJVV leaders attend these events. During fieldwork in 2012 there was an opportunity to attend a municipal assembly which focussed on the issue of metropolitan governance. The assembly was dominated by municipal staff making long speeches and there were few opportunities for CBO leaders to voice their concerns and responses. Despite limited opportunities for CBO leaders to participate, it was nevertheless interesting to observe how indigenous

JJVV leaders from Chasquipampa and Ovejuyo behaved and performed during this event:

> Francisco and Luciano seemed like completely different persons today. Normally Francisco wears a red poncho to official JJVV meetings and speaks in Aymara with Luciano and other members of the JJVV directive. Today, like everyone else in the audience, both men were wearing tailored suits with batges of the MSM [political party governing La Paz] tied to their jackets. Not once did I hear them say a word in Aymara. They only spoke in Spanish. They continuously praised the work of the municipality, something they never did in meetings in their neighbourhood, and expressed their support for the party and the issues discussed during the assembly.
>
> (Field-note diary entry by the author, November 2012)

Although Francisco and Luciano would wear different clothes and speak Aymara in their everyday lives in the neighbourhood, this no longer seemed of importance once they interacted with members of the municipality of La Paz. In this situation, they slipped into the shoes of different persons – they became *vecinos*. Francisco explained this behaviour as follows:

> You cannot wear a poncho when talking with the functionaries in La Paz. When I am in La Paz, I act as one of them, as a *mestizo*. In Palca, I am an authentic Aymara. To get the things we want from them we need to act and talk like them. This helps us develop friendships and friendships help you a lot in the municipality.
>
> (Interview, 19 November 2012)

Members from other JJVVs offered similar explanations. For example, Pedro – migrant and member of a JJVV in Chasquipampa – highlighted:

> I know how to talk to my neighbours and to the municipality. It's like two different worlds. Here we are Aymaras but there you cannot be an Aymara. Do you understand me? When I go to the municipality I learn to become one of them. I am very passive and respectful. I respect my engineers and my architects there. They receive me well in their offices. This helps me to bring back infrastructure works. For example, look outside my house here. The road to my house is now paved. This is good for my family but also for my neighbours.
>
> (Interview, 5 December 2012)

JJVV leaders, in other words, conform to officially recognised, static, and spatially fixed identity categories in order to pursue personal interests and the interests of some residents in their neighbourhood. Further, leaders' emphasis on friendship as well as on showing loyal support to local politicians and bureaucrats suggests that there exist clientelistic relationships between JJVV members and municipal staff. Jose, migrant and member of another JJVV in Chasquipampa whose community benefits from a large infrastructure improvement project called *barrios de verdad*, introduced by La Paz, explained this:

> I cannot complain about the work of the municipality. I have been a leader here for 14 years. I know all about their bureaucracy. By now, the architects in the municipality are my friends. For example, when I tell them that we need a road paved they will do this for our community. It was the same with *barrios de verdad* which is the project that was introduced by the old mayor Juan del Granada. I filled out all the forms correctly, prepared the application for this project very quickly and gave it to my friends in the municipality. Can you believe it? They gave *barrios de verdad* to my neighbourhood. Forty neighbourhood organisations applied for it and we got it. We got it simply because I am well respected in the municipality, have friends there and know their procedures.
>
> (Interview, 12 November 2012)

The issue of clientelism was further made explicit by the leader of the city-wide network of JJVVs – the FEJUVE. Even though the LPP defined JJVVs and FEJUVEs as politically independent units, La Paz's FEJUVE was directly linked to the political party *Movimiento Sin Miedo* (MSM) which holds a majority in the city council. The leader of La Paz's FEJUVE explained his connection to the MSM as follows:

> I have always supported the work of the MSM. They help us, we help them. There are some neighbourhood organisations that don't want to work with the MSM. They founded their own FEJUVE. Believe me, if you are with an opposition party, you don't get any services to your community.
>
> (Interview, 11 December 2012)

But JJVVs, regardless of whether or not they support the political party in control of La Paz's municipal government, did not help everyone. Indigenous residents who were not members of JJVV leadership boards often expressed a sense of mistrust towards their JJVVs. Women were particularly likely to say that they felt excluded from decision-making processes in JJVVs whose leaders were predominantly male. Another common criticism raised in conversations with residents was that while JJVV leaders manage to bring infrastructure services such as paved roads to the neighbourhood, these are often situated near their own homes and rarely benefit people who live elsewhere. Interventions by other CBOs, such as indigenous peasant unions, follow a similar pattern.

Indigenous peasant unions

In addition to accessing modern infrastructural amenities, *comuneros* also want to protect their remaining plots of collectively owned rural land for different purposes such as agricultural activities or speculation, that is, anticipation of selling the land in the future for a high price (see also Chapter 5). To protect or (re)claim collective lands, *comuneros* are organised in indigenous peasant unions which mainly negotiate with the municipality of Palca.

The present activities of indigenous peasant unions can be best understood by placing them in historical context (see Table 7.2). During the colonial and early postcolonial period, local peasants mainly worked under semi-feudal conditions, and land belonged to *hacienda* owners. Today's neighbourhoods of Chasquipampa and Ovejuyo belonged to the Patiño family and the Carmelita Sisters – a Catholic convent (Espinoza 2004). Following the 1952 agrarian reforms, land was redistributed to those who worked it – Chasquipampa's and Ovejuyo's indigenous *comuneros*. The agrarian reforms also defined the organisational structure for peasants in rural areas and, from 1952 onwards, *comuneros* organised in peasant unions linked to the National Confederation of Peasant Workers (CNTCB). In 1979 the Unified Syndical Confederation of Rural Workers – Tupac Katari (CSUTCB), founded by the indigenous Katarista movement – replaced the CNTCB and the neighbourhoods' peasant unions became members of a national union network that mobilised for both peasant and indigenous rights.

Table 7.2 Accessing collective land rights in Chasquipampa and Ovejuyo

Year	Regulation	Implementing government units and key practices	Tactics of indigenous residents
Until 1952	*Hacienda* system	National government	• Land belongs to *hacienda* owners • Indigenous peasants serve as landless semi-feudal peasants
1953	Agrarian Reform Decree (following the 1952 Bolivian Revolution)	National government abolished the *hacienda* system and distributed land to peasants	• Adopt class identity instead of indigenous identity when interacting with government authorities
2010	Law of Autonomies and Decentralisation (LAD 2010)	National government and municipality of Palca recognise individual and collective land rights of IOPs and grant IOPs access to specific constitutional rights	• Adopt role of 'indigenous original peasant' when interacting with government authorities

Source: Elaborated by the author.

Since the 1980s, the expansion of the city of La Paz impacted the area and most peasants sold parts of their plots to people who, subsequently, resold this land at a much higher price to incoming migrants. According to Pedro, a migrant from Chasquipampa, this generated anger among *comuneros*:

> They sold their land. They sold and sold and the city grew into this area. Soon they realised how much they lost in this process. Some of them sold their land for five Bolivianos [less than one US Dollar] per square metre. The person to whom they had sold it then sold it to a migrant for five or ten times the price. The *comuneros* were the losers in this process. At that moment, their peasant unions started to fight for the protection of their remaining land.
>
> (Interview, 5 December 2012)

Chasquipampa's and Ovejuyo's indigenous peasant unions relied on different tactics to protect their land. Members of unions often herded animals on undeveloped land demarcate their plot. A member of staff in La Paz's district municipality South (Interview, 27 November 2012) pointed out that members of the indigenous peasant union of Ovejuyo also used violence to protect their land:

> They attacked people as soon as they entered green spaces in the area. They attacked municipal staff as well. When we made our cadastral measurements some years ago they were beating us with sticks. They did all they could to avoid being registered by us.

In recent years *comunero* leaders have relied mainly on processes of political negotiation with Palca to protect their land. According to Roberto, a young *comunero* and member of the indigenous peasant union of Ovejuyo (Interview, 24 January 2013), there was always close cooperation with the rural municipality of Palca: 'Our union was always with Palca. We registered our land there. In Palca they understand us. In La Paz they want to destroy our fields. Why should we work with them?'

Depending on the political context, *comuneros* rely on different legal discourses and identity categories to justify their ownership of remaining plots of rural land. Carlos, a member of the indigenous peasant union in Ovejuyo, explained this as follows (Interview, 30 January 2013):

> We always avoided violent confrontation wherever possible. In the peasant union our brothers and sisters learned about the law. We always try to follow the law. First, it was the laws of the 1950s. Now it is the new laws of Evo.

Carlos also explained how his indigenous peasant union started to shift its negotiation strategy and how members 'became' IOPs after Evo Morales's

MAS was elected into national government and into the municipality of Palca:

> Comrade Evo says that the indigenous original peasants are the owners of their territory. The brothers and sisters in the municipality of Palca support this. They say that we are the indigenous original peasants here. We are writing a new constitution for our indigenous peasant union. Here we say that we, the original indigenous peasants of Ovejuyo, have the right to gain back our old territory. We want autonomy for our land.
>
> (Interview, 30 January 2013)

Comunero leaders generally relied on their ties with officials in government and national indigenous movements that are in line with Evo Morales's MAS party. This helped them to successfully negotiate demands for local autonomy and collective tenure rights. Francisco outlined this point as follows (Interview, 19 November 2012):

> In the city they are not with the indigenous people. They are *vecinos*. In the indigenous peasant union, we do not work with them in the city. We work with our people. We work with the CSUTCB, and with the municipality in Palca. They understand our concerns.

During meetings with the CSUTCB-TK and the municipality of Palca, *comunero* leaders directly referred to the Law of Autonomies (LAD) ratified by Evo Morales's MAS government in 2010 when articulating their claims (see Chapter 6 for further details). Besides a direct legal reference to new rights for IOPs, they also acted like IOPs in processes of political negotiations. In contrast to encounters with municipal staff from La Paz, where JJVV leaders would behave like *vecinos*, peasant union leaders – at times the same people who would take on leadership roles in JJVVs – would wear traditional hats and speak in Aymara when interacting with the municipality of Palca. But their behaviour apparently also changed over time. A member of staff in La Paz's municipality explained this as follows:

> Back in the old days they were the peasant class. They talked with the socialists to protect their land. Now the name has changed. With the MAS's rise to power, being indigenous is the new fashion. The old peasants are now indigenous original peasants. Their interests didn't change but their name and way of acting did.
>
> (Interview, 27 November 2012)

Folkloric associations

Indigenous residents also attach a strong cultural meaning to public places within their neighbourhood. The *Fiesta de la Virgen de Merced* is one example of cultural occupation of public space in the neighbourhoods (see Chapter 5). The municipality of La Paz does not authorise local residents to run the annual festival. A member of staff in La Paz's district municipality South explained why:

> The festival brings many problems. First of all, it is a security risk. They drink so much that things get completely out of control. We cannot allow this. Second, the festival takes place on roads built by our municipality. People often vomit on the streets and some spray graffiti. We do not want to pay for the clean-up, and for this reason we don't authorise the event.
>
> (Interview, 27 November 2012)

Some residents, particularly women, are aware of the security threat that the festival poses for their neighbourhoods. They associate the festival with a rise in violent behaviour and alcohol abuse. For example, four out of 12 focus groups conducted with Aymara women from Chasquipampa prioritised the issue of violence or insecurity as the main problem in the neighbourhood. In a causal-flow diagram, one focus group mentioned the annual festival as a security problem and highlighted alcohol abuse, aggressive behaviour, the absence of police, and the presence of too many people in the neighbourhood as the main causes of the problem (see Figure 7.1 for an illustration). To overcome problems of violence and insecurity, the women in this focus group shared the sentiments of the municipality of La Paz and suggested that the annual festival should be prohibited.

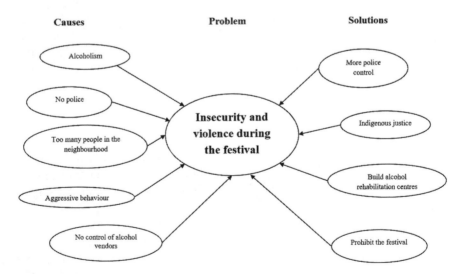

Figure 7.1 Causal flow diagram on the issue of insecurity in Chasquipampa.

Despite opposition from female residents and from the municipality of La Paz, the festive events take place year after year since folkloric associations such as the *Fraternidad Morenada Union Central Puma* manage to negotiate permits with the municipality of Palca and officials in the Ministry of Culture. *Comunero* Francisco described this process as follows:

> Back in the old days we always had problems with the police. This has changed since the MAS has been in government in Palca and in the national government. The mayor of Palca is an indigenous original peasant himself. He comes to our festival. He has plenty of friends in the Ministry of Culture who also support our festival. The municipality of La Paz complains about the festival, but they cannot do much about it. The national government is on our side.
>
> (Interview, 12 November 2012)

Legal authorisation from Palca and the national government not only allows *folkloristas* to run their festival but also prevents interference from the local police. Jose explains this as follows (Interview, 12 November 2012):

> The police are more accountable to the national government. The national government is controlled by the MAS. Palca is with the MAS. We are with the MAS and therefore the police won't interfere during the festival even if La Paz wants them to.

Youth tribal politics

In Chasquipampa and Ovejuyo, relations between indigenous leaders (who are predominantly male) and ordinary urban indigenous residents are also not straightforward. Rather, they are characterised by uneven power dynamics and conflicts of interest. Leaders play a crucial role in negotiating access to rights and services with national and local governments. Yet, taking advantage of their role as brokers between state and community, indigenous leaders often prioritise their own interests and those of their close friends while ignoring the interests of other members of their community, especially those of indigenous women and youth. This problem was mentioned by many younger residents in the area. The testimony of Ana Claudia, a young indigenous migrant from Chasquipampa, clearly outlines the problem:

> The old men think that they represent us but they really don't. All they care about is money, alcohol, old-fashioned festivals and football. For us young girls in the neighbourhood this normally just brings problems. When they are drunk, they want to kiss us and have sex with us. If we refuse, they threaten us. Many girls can tell you

their own experiences. To bring about long-term change, we have to follow our own pathways. That's why we founded our urban tribe. Here we educate ourselves, form future leaders who will do things differently, and connect with indigenous youngsters from across the country.

(Interview, 25 January 2013)

The youth tribe to which Ana Claudia referred was established by a group of ten college students, seven girls and three boys, in a local school in Chasquipampa in August 2012. In its initial stages the group would meet on the school's courtyard to run dance classes and rap shows. Mercedes, another member of the tribe and a rap singer, explained that these activities are not only recreational but political acts which are used for collective mobilisation:

Our music not only sounds cool but it also sends chilling messages. Our songs raise the problems of our neighbourhood – abuse of younger people by older ones, alcoholism, rape etc. When we started the shows many kids would come and listen and after a while we decided listening is not enough. We started running discussion groups and mobilised for our own space.

(Interview, 12 January 2013)

In December 2012, the founding members of the youth tribe managed to negotiate access to a room in a day-care centre which belongs to a local non-governmental organisation (NGO) operating in La Paz. The NGO staff connected this group to other youth groups in the city and enrolled them in human rights and social mobilisation training. Between 2013 and 2016, the youth tribe expanded its activities. It started to introduce fortnightly semi-nars and workshops in which young residents discussed among themselves the main problems and possible solutions for community development in their neighbourhood. Members of the youth tribe are now also part of a national platform of indigenous social communicators who, with financial support from different NGOs, meet yearly in a different city in Bolivia to exchange best practices. Felipe, member of the youth tribe, explained this as follows:

There is a new national movement emerging which is composed of a new generation of indigenous people living in cities who no longer care about who is migrant or *comunero*, Aymara or Quechua, man or woman, straight or gay etc. In our movement we want to depart from old hostilities between groups. Instead, we want to build unity in dis-advantaged communities. The first step to achieve this is to challenge outdated practices and confront our own old leaders.

(Interview, 23 November 2016)

And this is precisely what is starting to happen in Chasquipampa and Ove-juyo. To just give two examples: First, in 2014, the youth tribe set up its own arts committee whose responsibility it is to approach different organisa-tions, including La Paz's municipal unit for education and culture, as well as different local NGOs and arts foundations to request monetary or training support for the management and running of cultural events, such as con-certs, dance shows, or workshops in the neighbourhoods. Patricia, a mem-ber of the committee, explained that the role of the committee is not only to access resources but to disrupt patterns of social organisation within the neighbourhoods and to push for a more democratic and representational form of local community organisation:

> We formed the committee to change the way local politics works here. In the past, everything had to be channelled through the JJVVs but the leaders didn't give a damn about our concerns. We bypass them and approach people directly. We start with culture and hope to move on to other more contentious topics such as work, healthcare or housing. Unlike our indigenous leaders here who make decisions behind closed doors, we invite everyone interested to our assembly meetings and take decisions by public vote.
>
> (Interview, 24 November 2016)

Second, in 2015, the youth tribe established its own weekly radio programme where young indigenous residents are invited to share their own stories and problems. Raul, a young *comunero* from Ovejuyo, explained how the radio programme helped him to stir up resistance against illegal activities under-taken by an indigenous leader in his neighbourhood:

> When I was on the radio I spoke out against Franco, the leader of one of the indigenous peasant unions. I mentioned how Franco en-riches himself by illegally reclaiming land from people like myself and my family who purchased land from him decades ago. I asked our listeners to take to the streets and protest against him. The next day more than 200 people surrounded the peasant union and called for its closure. The protest seemed to work as Franco stepped outside and declared that he would resign and leave people in peace. And, indeed, he and no one else in the peasant union has dared to bother us since then.
>
> (Interview, 5 December 2016)

These examples clearly illustrate how the youth tribe changed from being a small collective of young women and men that mainly ran cultural events in a local college to a political platform that not only confronts unjust govern-ment practices but also challenges and condemns uneven power dynamics within urban indigenous communities.

Let us conclude this section by stating that in Chasquipampa and Ove-juyo residents not only articulate different interests, demands, and rights-based claims but also engage in distinct, and at times contradictory and conflicting, practices. Indigenous leaders play a key role in this process and, to get what they want, they slip in and out of different identity categories. These tactics of manoeuvring between different identity categories in different locales and at different times can be referred to as spatio-temporal identity politics. When and where to be or not to be indigenous in processes of political negotiations? This seems to be a central question for indigenous leaders. At one moment, leaders act as authentic indigenous persons when engaging with Palca and, at another moment, they hide their indigenous identity when interacting with La Paz. But leaders by no means represent everyone in the neighbourhood. It is for this reason that the underrepresented – particularly the youth – increasingly confront their leaders and transform their communities from within. This is done through the establishment of new CBOs, such as the indigenous youth tribe, which promote more democratic and inclusive organisational approaches and address the interests and needs of those who have historically been left behind not only by the state but also by their own indigenous communities.

Everyday indigenous urbanism in Quito

As in La Paz, Quito's indigenous migrants and *comuneros* voice their interests and rights-based claims through CBOs. *Comuneros* are mainly organised in *cabildos* (autonomous councils) which are increasingly engaging in collaborative planning processes with Quito's co-governance unit (see Chapter 6). Indigenous migrants are mainly organised in commercial vendor associations whose members often come from the same rural community. For example, according to a study conducted by Kingman (2012), within Quito's San Roque area alone, more than 30 indigenous migrant commercial associations represent indigenous migrants. It is these organisations, and particularly their leaders – who, as in La Paz, are predominantly male – that play a central part in negotiating with relevant government personnel over rights to preserve or access new work, community, and living spaces.

Given that most of Quito's indigenous peoples are migrants, this section focusses mainly on this group. It particularly focuses on the work of one CBO which represents indigenous migrants from the rural parish of Tigua – the AECT-Q. Over the last three decades, leaders of the AECT-Q have relied on a variety of self-help practices and engaged in processes of political negotiation with government actors in order to access spaces in which their members could work, interact, and live as a community. The remaining parts of this chapter offer three case studies which illustrate in detail the practices of the AECT-Q.

Accessing collective work spaces

The majority of indigenous migrants who came to Quito from the rural parish of Tigua worked in Quito's San Roque market which, at the time of the author's fieldwork in 2013, was facing closure. Being confronted with threats to relocate from their work places in San Roque was nothing new for members and leaders of the AECT-Q. Indeed, city-centre revitalisation efforts had already been planned by the former mayor of Quito, Paco Moncayo, and his Democratic Left party which held a majority in the city council from 2000 until 2009. This municipal government drafted a city-centre plan in which it outlined a set of interventions to make Quito more attractive for international tourists (DMQ 2003). For example, to increase safety and security within the city centre the municipal government attempted to prohibit street vendors from selling their wares in this part of the city for a detailed discussion on this topic see also: Swanson 2007). Confronted by relocation and displacement threats, indigenous leaders from the AECT-Q relied on a diverse set of tactics to defend the interests of their members. According to Raul, former leader of the organisation and a teacher in the Chaquiñán College, the AECT-Q articulated their demands to preserve or gain access to new work spaces in such a way as to ensure that they were in line with the municipal agenda on promoting tourism:

> Paco Moncayo wanted to bring more tourists to Quito. This helped the Tiguan painters a lot. We went to the municipality and showed them our paintings. We explained to them that the tourists come to Ecuador to see our folkloric art. We told them that we, the indigenous people from Tigua, are a tourist attraction. This worked and from that point the municipality did not make problems for us anymore but offered us a permit to sell our products in the park. Until today our brothers and sisters who work as painters can sell their products there.
>
> (Interview, 27 April 2013)

Tiguan painters managed to gain access to new commercial spaces in the city's El Ejido Park which is located between the city centre and Mariscal Sucre – an upmarket neighbourhood which is home to many government offices, businesses and tourist facilities. While painters could mobilise around Quito's municipal tourism agenda, such a negotiation approach did not work for Tiguans who traded products such as vegetables and fruit in the San Roque market or on the streets of the city centre. According to former AECT-Q leader Rodrigo (Interview, 26 April 2013), civil servants in the municipal government refused to change their decision to relocate indigenous vegetable or fruit vendors whom they often described as an 'eyesore' for tourists. Such a negative perception of indigenous market vendors has also been reported in other studies. Mary Weismantel (2001), for example, argues that municipal authorities across Ecuador have historically viewed urban markets as indigenous and, hence, messy and dangerous places that disturb the structure of the 'white' and 'orderly' city.

In a context where political negotiations did not produce positive results for all market vendors, the leaders of the AECT-Q had to find alternative solutions – they relied on their connections to other indigenous market vendor organisations who confronted similar problems. In October 2001 the AECT-Q, in alliance with 14 indigenous market associations, founded the organisation *Jatun Ayllu* which represented more than 3,000 San Roque-based indigenous market vendors. *Jatun Ayllu* was directly affiliated with the national indigenous movement CONAIE and the political party Pachakutik. With help from CONAIE leaders and other indigenous organisations, such as CODENPE, *Jatun Ayllu* developed a plan to construct a new indigenous commercial centre which could provide space for all its members. *Jatun Ayllu* proposed to the municipal government of Quito that it would finance the construction of the commercial centre with money donated by its members and through financial support from CODENPE. In exchange, the organisation demanded from the municipal government receipt of a plot of land. In 2008, Quito's municipal government did, indeed, provide *Jatun Ayllu* with a 41,000 square-metre plot of land in the Chillogallo district situated in Southern Quito. The municipal government also guaranteed not to displace indigenous migrants from San Roque until the new commercial centre had been built.

The dream of a new commercial centre for all indigenous migrants from San Roque was, however, a short one. In 2010 Quito's municipal government, which at that point was led by mayor Augusto Barrera, reclaimed the 41,000 square metres of land, declined building permission, and stopped processes of political negotiation with *Jatun Ayllu*. A former member of *Jatun Ayllu*, Orlando, mentioned the following reasons for the abrupt end of the commercial centre project:

> *Jatun Ayllu* was with CONAIE. Being with CONAIE was helpful as different people in government would open their doors for us. At present, CONAIE opposes [Rafael] Correa's party. Everyone who is with the opposition struggles to receive anything from the government. When Augusto Barrerra and the Alianza Pais got elected here in Quito it was over for us. No more *Jatun Ayllu* and no more commercial centre.
>
> (Interview, 10 April 2013)

According to this interpretation, the end of the commercial centre project was a political manoeuvre by a newly elected municipal government, which, like Ecuador's current national government, was interested in disempowering opposition groups like the country's indigenous movement (see Chapter 6). People working in Quito's municipal government have confirmed such tendencies. For example, a civil servant working for the municipal commercial unit during Barrera's administration explained the following:

We do not work with *Jatun Ayllu* because they support CONAIE, Pachakutik and all of the other social organisations that want to cause social
unrest in this country. They are corrupt, internally divided, and don't
respect their base. Why should we still set out to help them to build their
commercial centre? We work for the people that elected us. We do not
work for the opposition.

(Interview, 10 July 2013)

Internal problems such as corruption within *Jatun Ayllu* were also mentioned by indigenous migrants themselves and represented another cause
for the failure of the commercial centre project. Juan Carlos, a member of
AECT-Q, explained this as follows:

Our brothers and sisters paid money to *Jatun Ayllu* so that they would
build the commercial centre. It was clear that most of the money went
not into the commercial centre but into the pockets of *ponchos dorados*
[rich indigenous leaders]. With no commercial centre in sight and our
money gone, we decided to leave *Jatun Ayllu*. Now we work only for the
Tiguans again.

(Interview, 10 May 2013)

Accusations of corruption were also raised by other indigenous migrants
from Tigua approached in the Chaquiñán College who mentioned that they
had paid *Jatun Ayllu* leaders the equivalent of 150 US Dollars to secure themselves a stall in the new commercial centre. Three years after the official can-
cellation of the project, however, their money had not been returned to them.

In such a context it was not surprising that indigenous migrants no longer
trusted the state and civil society organisations that were supposed to
represent their interests. Such trends were, for example, made explicit in
institutional maps drawn by focus groups in the Chaquiñán College (see
Figure 7.2). In this institutional map, the focus-group members evaluated
the work of actors and institutions associated with the government, such as
Rafael Correa, the Ministry of Education, and Quito's municipal government, but also indigenous movements and CBOs, such as CONAIE, ECUA-
RUNARI, and *Jatun Ayllu*, as very negative.

The members of this focus group perceived their own indigenous market
vendor association – the AECT-Q – and international NGOs such as Plan
International more positively. Plan International represents an important
organisation in Ecuador which facilitates urban development at the community level (Moser 2009). At the time of the author's fieldwork in Quito,
Plan International was working in the San Roque area on issues such as
child labour and the training of street vendors targeting mainly vulnerable
adolescents and young adults. It provided them with access to free day-care
facilities and educational workshops. Most focus-group members attended
some of Plan International's activities and evaluated the organisation

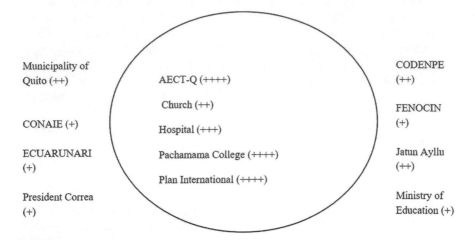

Municipality of
Quito (++)

CONAIE (+)

ECUARUNARI
(+)

President Correa
(+)

AECT-Q (++++)

Church (++)

Hospital (+++)

Pachamama College (++++)

Plan International (++++)

CODENPE
(++)

FENOCIN
(+)

Jatun Ayllu
(++)

Ministry of
Education (+)

Figure 7.2 Institutional map of the San Roque market area.

positively (Field-note diary entry, 28 April 2013). The positive perception of the AECT-Q can most likely be explained by the organisation's ongoing effort to unite Tiguan vendors within one commercial, community, and living space (see also the next two sections).

To continue their work in the current political environment, the leaders of the AECT-Q decided to leave *Jatun Ayllu* and continued working only for their Tiguan migrant base. Yet, for an organisation whose leaders were affiliated with CONAIE, entering into political negotiations with the AP-led municipal government continued to be difficult. The AECT-Q member Juan Carlos explained this as follows:

> We want to be reallocated all to one market in order to continue working and living as a community. Barrera's municipality does not understand this. When working with us vendors they no longer recognise our association. They believe we are still with *Jatun Ayllu* and with the CONAIE. Instead of talking to our leaders, they work with each of our members individually. This might mean that each one of us will be allocated to a different place and we won't be treated as a community.
>
> (Interview, 10 May 2013)

Lacking political allies within Quito's municipal government, indigenous leaders could influence decision-making processes only by preparing their community members for individual meetings with officials in the commercial unit. Ricardo, an AECT-Q leader, described this as follows:

> The municipality asks each vendor to identify their preferred work location and, after assessing these places, they might relocate them there.

We tell all our vendors to name the Mayorista market. If we are lucky we will all receive a space in this market. In the current political situation there is not much else what we can do.

(Interview, 11 May 2013)

Relocation has so far not occurred. In fact, the relationship between the AECT-Q and Quito's municipal government changed in a positive direction after the election of a new mayor in 2014. As was already outlined in Chapter 6, Quito's current municipal government, which stands in opposition to the political party Alianza Pais (AP) drives a more pro-indigenous agenda, and has departed from a relocation approach in San Roque. Instead, leaders of indigenous organisations such as the AECT-Q are being consulted about plans to incrementally upgrade the San Roque market and surrounding streets. While AECT-Q leaders perceive this trend as positive, they also mention that consultation takes place only as long as they support the current municipal government. Ricardo explained this as follows (Interview, 20 October 2016):

The game we play has not changed. To be heard, we have to pretend to be their supporters. This helps us at the moment, but who knows who will govern in a few years' time? The struggle must continue but it is always uncertain what will happen next.

Struggling to preserve community: the case of the Chaquiñán College

To make a living in the city, indigenous migrants often work long days and nights, lack time for other activities such as accessing education. Leaders of the Tiguan migrant community were aware of this problem and found ways of educating the members of their migrant community. Indigenous leaders who benefitted from secondary education started to provide open-market classes in Spanish, Kichwa, and maths to Tiguan migrants. To improve their pedagogical skills, they volunteered every weekend in a bilingual school which was situated in the Cotopaxi province near the rural parish of Tigua. Following the reforms on intercultural education in the 1980s, this bilingual school was administered through the National Directorate of Intercultural Bilingual Education (DINEIB) and was run entirely by indigenous for indigenous people. Through volunteering in this school, indigenous leaders learned about intercultural education and this allowed them to introduce the same curriculum and pedagogical principles to their urban community in Quito.

With the help of acquaintances in DINEIB and the CONAIE, the AECT-Q managed to negotiate access to a vacant building near the San Roque market in which they opened the Chaquiñán College in 1994. Similar to previous open-market classes, the Chaquiñán College was managed entirely by volunteer teachers who were associated to the AECT-Q. In contrast to conventional schools, the Chaquiñán College provided classes to indigenous migrants of all ages. Most of the students were young adults who, having received no education during childhood, aspired to receive a school diploma at a later stage in life. The school operated only on Saturdays so that students and teachers could work full-time throughout the week. By being recognised as an official extension of the school in the Cotopaxi province, the Chaquiñán College received accreditation from DINEIB as a registered bilingual education institution in 1995 and was allowed to award students with secondary education certificates. The premises of the Chaquiñán College were used not only for purposes of education but also for association meetings of the AECT-Q and for the realisation of a variety of community events such as the *Inti Reymi*.

After operating for 17 years as a formally registered weekend school recognised by DINEIB, the director of the Chaquiñán College was notified in January 2013 by Ecuador's Ministry of Education that the school must stop its activities by the end of the academic year in 2014. The letter specified that, according to Article 13 of the new Law on Intercultural Education (LIE), bilingual weekend schools which relied on voluntary teachers were no longer allowed to operate in neighbourhoods in which other registered schools provided intercultural education to students on weekdays (see also Chapter 6). For the following academic year, the students of the Chaquiñán College should, therefore, be relocated to nearby secondary schools, while the premises of the Chaquiñán College must be returned to its owner – the municipal government of Quito.

The reasons for the school's closure were certainly legitimate. Voluntary teachers often lacked university education and were not certified by the Ministry of Education. Moreover, there were four normal schools in the San Roque neighbourhood that could absorb some of the students. But, while the letter provided legal justification for the school's closure, many teachers and students of the Chaquiñán College argued that the real reason was that the municipal government intended to renovate the school building before selling it to private investors. Such intentions were indeed confirmed by municipal staff. A former director of Quito's territorial planning unit working in the Barrera administration, for example, stated:

> As with other buildings in the neighbourhood such as the ex-prison or the market itself, we intend to renovate the Chaquiñán College building. This neighbourhood has a lot of potential. It is right next to the colonial city centre. A building like this could be a new cultural centre, a hotel, an apartment complex. We have not established any further plans for this building at this moment but, yes, it will no longer belong to market-vendor associations in the future.
>
> (Interview, 22 May 2013)

As a result of the city-centre revitalisation project, the members of the AECT-Q would lose not only their work spaces but also their central community space. Teacher Juan Carlos explained this:

> The closure will be devastating. Who can teach our children about our community better than we people from the community? Our school is the centre of our community, and taking it away will weaken us. The closure of the school will stop our association members from learning and working at the same time. Our brothers and sisters will become poorer.
>
> (Interview, 10 May 2013)

Like Juan Carlos, many students associated the closure of the school and their potential integration into normal schools which operated throughout the week with a decline in their personal income. Adult students in particular had to abandon their education as they simply could not give up their job in the market throughout the week. Parents were also concerned about sending their children and adolescents to schools that operated only on weekdays. Even though they could apply for the government's new *Bono de Desarrollo Humano* which provided them with a monthly stipend of 35 US Dollars in return for sending their child to school, they often argued that they would earn more money with their child working in the market.

Confronted by the imminent closure of their school, AECT-Q leaders swiftly engaged in the search for new community spaces so that they could continue their activities elsewhere. They did this by entering into political negotiation with the authorities in Quito's municipal government. The teacher Raul highlighted this as follows:

> Before Correa we could take to the streets and fight for our rights as indigenous peoples. Remember all the marches by the CONAIE? Well, we were always participating. You know we are part of the CONAIE. The current government has criminalised indigenous protest and the CONAIE. They would put you in prison if you went on the streets with them. Today, under the government of Correa, it is no longer a fight which we carry out on the streets but one which we can win only through words. If we want them to give us a new space for our school, however, we must be careful with the words we use. If you want to achieve something you must be with the government and not with those that oppose them. You must go to those in the government and speak their language.
>
> (Interview, 27 April 2013)

Indigenous leaders like Raul, who were also involved in negotiating access to new market spaces, learned that, in the current political context, their affiliation to indigenous movements such as CONAIE hindered them in addressing their specific demands. They therefore started to adjust their negotiation tactics and became active supporters of the political agenda which they had previously opposed. This was evident during meetings between leaders and municipal authorities which normally took place in a community development centre (CDC) in San Roque. For example, during one meeting AECT-Q leaders openly articulated their support for the government's decision to close their own school. They highlighted that as supporters of Correa's AP party, they would appreciate gaining access to the CDC to continue their cultural and educational activities. To prove their support for the AP, these leaders showed photos of their members wearing AP shirts and participating in the annual Labour Day march. And municipal staff granted the AECT-Q permission to use the CDC once a month for two hours in order to undertake cultural events. However, they did not permit AECT-Q leaders to continue their educational activities (Fieldwork diary, 25 May 2013).

According to leader Rodrigo, the AECT-Q continues to make strategic use of vote-bank politics and supports for whoever is in office in order to preserve the few remaining community spaces available to them in the city:

> When Barrera's administration left, we quickly adjusted. We needed new T-shirts to show our support for the SUMA party and Rodas. And, so far, it has at least helped us gain access to the cultural centre every couple of weeks.
>
> (Interview, 20 October 2016)

Accessing collective living spaces

Besides aspirations to work and socialise in one collective space, indigenous migrants often highlighted that they wanted to live in the same neighbourhood (see Chapter 6). At the time of my fieldwork in Quito in 2013, indigenous associations such as the AECT-Q still rented colonial houses in the San Roque area but was unable to provide a living space for every association member. Housing was a problem not only for indigenous migrants but also for a large number of Quito's population. In 2012, 132,461 residents lacked access to adequate housing in Quito (DMQ 2012b). Quito's municipal government, during the term of mayor Augusto Barrera, sought to target the housing problem through the implementation of large-scale social-housing projects. In addition, it promotes the creation of resident initiatives to design and implement their own housing projects. In exchange, the municipal government provided these resident initiatives with financial and technical support as well as with access to cheap vacant public land within the city's periphery.

Aware of the housing needs of their members, AECT-Q leaders make use of the municipal governments' approach to residential housing initiatives.

Since 2013 and until the present, leaders of the AECT-Q and also of other indigenous organisations – such as the AVIC-Q (Association of Indigenous Vendors from Cotopaxi who reside in Quito) and Intimama – are in the process of designing and implementing their own housing projects in which they guarantee each of their association members a plot of land to build their own house (see Table 7.3). In order to implement their housing projects, the leaders of the three organisations negotiated access to a plot of land and building permissions with the municipal government of Quito. They all received support from an architect who was previously employed by the municipal government of Quito. According to information received from this architect and from leaders of the associations, by April 2016 the three housing initiatives were at different stages. Intimama had received a plot of land and relevant building permits. The members of this association had finalised the construction of 65 houses and managed

Table 7.3 Housing projects of indigenous migrant associations

Association	Membership basis	Housing project	Expected residents	Interactions with municipality
AECT-Q	Indigenous migrants from Tigua who work as market vendors in Quito's San Roque and Mayorista markets	Build up to 150 houses and a community centre in Quito's Guamani district	800	Received a plot of land from the municipality of Quito Awaiting building permission
AVIC-Q	Indigenous migrants from Tigua and elsewhere in the Cotopaxi province who predominantly worked in the San Roque and Mayorista markets	Build up to 400 houses and a community centre in Quito's Guanmani district	2000	Applied for land and building permission
Intimama	Indigenous migrants from the Chimborazo province who predominantly work as market vendors and in commercial centres in Quito's city centre	Build 65 houses in Quito's Quitumbe district Space for a community centre, a cooperative bank and a sports ground	200	Received a plot of land and building permissions from the municipality of Quito In process of applying for access to public services (water, electricity, roads etc.)

Source: Elaborated by the author.

to gain access to public services, such as water, electricity, and roads. The AECT-Q had received plot of land in Quito's Guanmani neighbourhood in 2012 and been granted the equivalent of USD 100,000 of financial support from Ecuador's Ministry of Urban Development and Housing (MIDUVI) to cover two-thirds of the costs of this land. By 2016, construction had begun but no one yet inhabited the area. AVIC-Q, meanwhile, is still in the process of applying to receive access to a plot of land in Quito's Guanmani district.

Despite being at different stages with their housing projects, the leaders of these three organisations share the fact that they rely on similar political negotiation tactics when interacting with municipal government officials. Orlando, the leader of the association AVIC-Q, provided a clear illustration of the negotiation approach of indigenous leaders:

> In the commercial unit they all knew me and didn't want to see me. They accused me of being of the opposition because I used to be with CONAIE and *Jatun Ayllu*. 'Go away, *indio sucio* [dirty 'indian']' they said. The people in the municipal housing enterprise did not know me. I told them that I represent indigenous people who live in bad conditions in San Roque and elsewhere in the city. I told them that we want to get better housing. I also told them that I and the members of AVIC-Q were with the government. I showed them pictures of how we supported the government during marches. They liked this and in response they looked at my project proposal and allowed me to apply to get some vacant land in Guanmani. This is how it works here in Quito – we do something for them and they do something for us.
>
> (Interview, 10 April 2013)

A similar explanation was provided by Juan, a leader of AECT-Q:

> I would tell no one in the housing unit that I am with *Jatun Ayllu*. I tell them that I am with the government and want to help the authorities to solve their relocation problem. I bring the project and they can give me cheap land. This helps me to get on well with the people there.
>
> (Interview, 8 May 2013)

Once again, then, it seems that the ability to manoeuvre between different political agendas determines the success of indigenous leaders in political negotiations. This was made clear by another leader of the AECT-Q, Juan Carlos:

> I have learned from my mistakes. In the past we used to be with one political group but we soon learned that you don't get anything with this. Our governments change from one day to the next here in Ecuador.

You need to become flexible. You should never show whom you really support. To achieve the goals of our organisation you have to work with every government and support each of their campaigns. This means that sometimes we need to be with the indigenous movement and sometimes not. What matters is that you have good relations with all of them.

(Interview, 10 May 2013)

In the context of housing, indigenous leaders not only have to manoeuvre between being supporters of distinct political agendas. They also have to present their housing projects in such a way that they comply with the specific demands of their association members. The AVIQ-Q leader Orlando explained this as follows:

The municipality does not care about our community values. All they want is that we leave the city centre. We need to show our support to get things done. It is my brothers and sisters in the association that want to live in a community. I will have to make sure that I cater for their interests as well when I plan our community.

(Interview, 10 April 2013)

When working with members of their own associations, indigenous leaders generally presented their housing projects as a way of reinventing the indigenous community. During community meetings, for example, the AECT-Q leader Juan promised the members of his association to construct a new community centre on the plot in Guamani where, after the closure of the Chaquiñán College, they could once again revive educational, cultural, social, and political activities.

But indigenous leaders also intended to benefit personally from these housing projects. For example, AECT-Q's leader, Juan, made this point as follows:

This is my project. I made the connections to the government. I gathered some financial support. I have always represented my community and their needs. I helped them to save money and now I can give them financial support so that they can pay for their houses. All this takes a lot of my time and effort. In exchange for all this, I expect a little contribution from the members of AECT-Q. I give them credit but, like the other banks, I charge them an interest rate. This is how I do business.

(Interview, 8 May 2013)

Indeed, Juan provides association members with microcredits, normally with a 25 per cent interest rate, so that they can pay for their new plot of land. AECT-Q members often mentioned that they increasingly dis-trusted Juan since he failed to provide them with evidence of the actual cost of the land or a certificate stating that they purchased the plot. In addition, when I returned to Quito in 2015 and 2016, AECT-Q members said that Juan had started to sell plots of land to people outside their community. Juan confirmed this trend but explained that, in a context in which some members cannot pay for their allocated plots, he is left with no other option than selling vacant plots to other people who are often willing to pay him an even larger sum of money. Indigenous leaders like Juan, therefore, seem no longer only involved in the creation of new ur-ban indigenous communities but engage in activities of land speculation for personal profit.

In summary, then, efforts to access collective work, recreational and living spaces are characterised by two clear patterns. First, it is pre-dominantly indigenous leaders who raise specific interests, needs, and rights-based claims by means of political negotiations with government authorities. Second, when interacting with government authorities, lead-ers modify their negotiation tactics over time in such a way that they express their demands in line with the specific political agenda of the government which is currently in power. It is this ability to manoeuvre between different political agendas which allows indigenous leaders to address their own personal interests as well as the interests of the people whom they represent.

Conclusion

This chapter has demonstrated that indigenous residents are not passive actors but active planners of their own lives who seek to fulfil their inter-ests and demands and, in so doing, claim their indigenous right to the city. In the current political context, indigenous peoples in La Paz and Quito mainly make use of invited spaces (Miraftab 2009). They approach gov-ernment authorities to engage in political negotiations or participatory processes. But not every urban indigenous resident engages equally in the struggle for the indigenous right to the city. Rather, the findings presented in this chapter emphasise that historically marginalised indigenous com-munities are themselves characterised by uneven power relations and that it is community leaders who obtain the relevant social and political re-sources to negotiate with government authorities and engage in participa-tion processes. In both cities, the majority of indigenous leaders are older men. This reflects a general trend in Latin America, where community organisation follows principles of *machismo* and *marianismo* according to which men often take a stronger political position within the community than women, whose role is restricted to domestic affairs and care work (Moser 2009).

Two further findings stand out from the discussion of the different practices of indigenous leaders in La Paz and Quito. These relate, first, to how they act in processes of political negotiations and participation and, second, to whom they represent through such practices. In both cities, indigenous leaders must be capable of performing different roles and rely on plural political tactics in order to successfully articulate their interests and rights-based claims. But evidence from La Paz and Quito also suggests that the tactics of indigenous leaders vary between cities and change over time and across space. In La Paz, indigenous leaders have to approach different local government authorities to discuss access to services and recognition of different rights. In this process, leaders conform to multiple yet very different spatialised identity categories. When interacting with Palca they perform as authentic and traditional indigenous subjects, while in negotiations with La Paz they have to become 'white' residents of the modern city. In contrast, indigenous leaders in Quito shifted over time from being indigenous-movement affiliates to being supporters of a political party which perceived indigenous movements as their opposition. Through such tactics and through constant manoeuvring between different political environments, indigenous leaders not only get what they want, but they also manage to adjust the existing political rules and procedures to their own interests and to the interests of those whom they represent. As such, indigenous leaders define what the indigenous right to the city actually means on the ground in different local settings.

The findings presented here also reveal that indigenous leaders do not always address the interests and demands of all their CBO members or of their communities. Instead, they often use their powerful positions to enrich themselves personally or to provide close friends with access to resources while denying such access to other community members such as women or younger residents. Once again, this speaks against a romantic and essentialised interpretation of urban indigenous communities as harmonious entities. Instead, urban indigenous communities – as with most urban communities elsewhere in the world (Banks 2014; DeFilippis 2001) – should be considered as divided places characterised by their own internal hierarchies, conflicts of interest, and unequal power relationships.

Finally, the chapter demonstrated that exclusionary practices within indigenous communities no longer remain unchallenged. This was illustrated through a discussion of the practices of the indigenous youth tribe which operates in La Paz's Chasquipampa and Ovejuyo neighbourhoods. The work of the youth tribe highlights that a more inclusive and pro-indigenous politics must not only confront state-led practices which reinforce patterns of ethno-racial exclusion. It is equally important to problematise patriarchy, intergenerational conflict, and uneven power relations within urban indigenous communities, factors which contribute to the trend that indigenous women and youngsters remain at the bottom of the ethno-racial hierarchy. More attention needs to be paid to such intra-community problems and to strengthening the role of the marginalised within historically marginalised communities – indigenous women and youth – so that their voices will be heard and their interests considered in processes of translating the indigenous right to the city at the level of communities, of cities, of nations, and internationally.

8   Conclusion

This book has examined the role of indigeneity in urban policy and planning practice, and explored struggles for indigenous rights in cities – places which are conventionally conceived of as 'non-indigenous.' In Latin America, in particular, the term 'indigenous city' reads like an oxymoron. Chapter 2 highlighted how, following the colonial conquest, the city was conceived of as non-indigenous space associated with 'white' Spaniards or people of 'mixed blood,' with 'modern' infrastructure, such as the town hall, cathedral, or road grid, and citizenship as well as individual property rights. In contrast, the countryside was conceived of as an 'Indian,' 'native,' or indigenous place, as space which lacks 'modern' infrastructure but is characterised by 'traditional' forms of community governance. Such ethno-racial, spatial, political, social, and economic hierarchies were often reproduced by postcolonial regimes, leading to a situation of coloniality.

Challenging this discourse, this book has demonstrated how indigenous peoples were never fully excluded from urban life. Since the second half of the 20th century, in particular, indigenous peoples have increasingly moved to cities and engaged in a variety of bottom-up urban political struggles to voice their specific interests and demands. These struggles were defined in this book as 'indigenous rights to the city' – a cry and demand to be recognised as urban indigenous residents with distinct interests, and to be involved in urban decision-making processes. In Bolivia and Ecuador, urban indigenous uprisings led to the ousting of governments and to the election of new governments which, for the first time in history, recognise indigenous rights to the city in national constitutions.

The aforementioned constitutional reforms represented the analytical starting point for this book, which explored the translation of indigenous rights to the city into urban policy and planning practice. This process of translation was conceptualised as involving a multiplicity of social actors – officials working in government institutions associated with urban governance but also urban indigenous residents and their community-based organisations (CBOs) (Chapter 3). Relying on a qualitative methodology and case-study comparison, the book examined the translation of indigenous rights to the city by focussing on two case-study cities – La Paz, Bolivia, and

Quito, Ecuador (Chapter 4). Through deploying a variation-finding comparative method, the book revealed that the translation of indigenous rights to the city is influenced by different causal factors within distinct urban settings, including cities' unique histories and specific local political and institutional structures. Hence, variations in the findings *between* the case-study cities were explained through shedding light on the unique processes and factors that shaped the translation of indigenous rights *within* each city.

The findings presented in this book are mixed. On the one hand, the official revaluing of urban indigeneity and the recognition of indigenous rights to the city have had immense positive effects in La Paz and Quito. In La Paz, residents mentioned how having a President, Evo Morales, who himself identifies as Aymara has not only boosted their pride in being indigenous but also provided them with new opportunities. As outlined in Chapter 5, indigenous residents are increasingly holding positions in government institutions and in the business sector which were previously reserved for 'white' and *mestizo* elites. This had important class effects, allowing some residents to become part of a new Aymara bourgeoisie. In both cities, indigenous residents also highlighted their satisfaction with government redistribution policies, such as pension schemes or school bursaries, which contributed to a reduction in poverty levels and socio-economic inequalities between different ethno-racial groups. These are very positive accomplishments and represent an important historical transformation of ethno-racial relations within urban Bolivia and Ecuador.

On the other hand, though, understandings of indigenous rights to the city – whether in the realm of everyday life (Chapter 5), policy and planning (Chapter 6), or bottom-up political struggles (Chapter 7) – were characterised by a set of conflicting realities. The remaining sections of this concluding chapter focus on these conflicting realities and offer some broader reflections for engaged scholarship and practice on topics such as indigeneity, marginalised 'communities,' and alternatives to urban development such as those framed around the notion of the right to the city.

Urban indigeneity as lived experience: embracing heterogeneity and difference

Spatio-temporal static understandings of indigeneity continue to dominate scholarly and policy debates on indigenous peoples. This is evident in Bolivia where, despite the recognition of indigenous rights to the city in the constitution, the authentic indigenous person is considered to be an 'indigenous original peasant' or, in other words, a person living in remote rural areas or isolated forests. In principle, only people who fit this definition are granted with collective indigenous rights (see Chapters 2 and 6). Similar trends can also be noted in international campaigns which continue to portray indigenous peoples as 'guardians of the forest.' Such representations offer an essentialising, romantic, and ruralist account of indigenous peoples and thereby explicitly ignore the urban indigenous majority that lives outside forests in concrete jungles, including in peri-urban neighbourhoods such as Chasquipampa and Ovejuyo in La Paz or in *communes* and migrant communities in Quito. In these urban settings, different indigenous residents aspire to make the best of different worlds. They want to lead a modern life in the city while preserving or revitalising their traditions. They claim universal and individual rights but also seek to gain recognition for collective indigenous rights. It is perhaps this combination – between traditional and modern, and individual and collective – which defines contemporary urban indigeneity.

The book demonstrated how an asset accumulation framework can capture the differentiated, conflicting, and constantly shifting articulations of urban indigeneity. The use of an asset accumulation framework revealed that, in La Paz and Quito, being indigenous in the city means very different things to different people. In both cities, interests and demands varied depending on origin (e.g. migrant or *comunero*), age, gender, and political position. The only thing indigenous residents shared in common was that they articulated their interests and demands in relation to land. For example, indigenous residents referred to land or public space when raising conflicting demands to access cultural (e.g. traditional *fiestas* vs modern rap events), financial (e.g. money generated from reselling and subdividing collective land), physical (e.g. access to urban amenities, such as water, electricity, or roads, vs access to agricultural land), natural/productive (e.g. for agricultural activities vs urban speculation), or politico-legal (e.g. collective vs individual rights) goods and resources.

To broaden our understanding of what urban indigeneity means to people living in different settings and moments in time, future research could make use of the asset accumulation framework introduced in this book and apply it to other urban settings in the global South or North where a growing number of indigenous peoples now live. This will further contribute to efforts to de-essentialise and urbanise the topic of indigeneity as well as shed light on the multiple and diverse meanings of what it means to be indigenous in an increasingly urbanising world.

From discourse to implementation: urban governance and indigenous rights to the city

This book challenges generalisations made in previous studies which claim that the interests and demands of indigenous peoples cannot be met as long as different actors involved in urban governance follow liberal approaches to urban development and planning, and prioritise individual rights over specific group rights (Chapter 3). As highlighted earlier, most urban indigenous residents, at least in La Paz and Quito, want to receive individual tenure rights in order to access basic public services, such as water, electricity, or road infrastructure. In this regard, they are not that different from other urban residents who want to benefit from the modern amenities of urban life. Planning models which follow individual rights-based approaches should therefore not be automatically disregarded as they do yield results which are responsive to the interests and demands of many urban indigenous residents.

The book also demonstrated that indigenous rights to the city are unlikely to be translated into policies and planning interventions in contexts in which public officials responsible for their implementation continue to hold preconceived views of cities as non-indigenous spaces, follow different political priorities, or operate in a political environment characterised by conflicts between different local authorities (see Chapter 6). However, the book also discussed examples of national and local government interventions which tackle the different interests and demands of urban indigenous residents and, therefore, offer some ideas on how state interventions can positively embrace indigenous rights to the city in all their diversity. Best practices identified in this book were, among others:

- Intercultural education reforms introduced by Bolivia's and Ecuador's national governments.
- Efforts to design and implement an alternative pro-indigenous urban plan by La Paz's intercultural unit.
- Attempts by Quito's co-government unit to embrace conflicts between indigenous communes and local government authorities through integrating municipal and indigenous governance schemes as well as individual and collective tenure rights (see Chapter 6).

Interviews with policymakers involved in the aforementioned interventions highlighted that current practice remains far from perfect. But these policymakers also suggested ideas for improvement. For example, for intercultural education reforms to work on the ground it would be necessary to confront and alter racist attitudes which continue to guide the everyday practices of local teachers. Meanwhile, La Paz's intercultural unit and Quito's co-government require better institutional support as well as financial and human resources to implement their ambitious plans and policies.

In summary, the findings from this book suggest that the state is by no means monolithic. Instead, different social actors operating within the state offer different responses towards the translation of indigenous rights in cities, with some yielding more positive results than others. The findings also suggest that it is, perhaps, better to depart from *a priori* assumptions or ideal-type models on 'what could be done' to address urban indigenous peoples. Instead, it is important to identify what works best in specific local settings. Rather than conclude with a set of universal policy recommendations, then, the author calls for further in-depth and empirically grounded research which captures the perceptions, interests, views, and associated practices of diverse social actors involved in urban governance, who operate in different cities characterised by their own institutional, political, and structural environments. In doing so, it is possible to identify local and context-specific pathways for the successful translation of indigenous rights to the city.

Claiming indigenous rights to the city from below: disentangling 'community'

The findings presented in this book speak against an understanding of indigenous communities as harmonious and egalitarian collective entities. They also speak against interpretations of the right to the city as a 'collective' cry and demand articulated by the marginalised and excluded. Instead, this book revealed how historically marginalised indigenous 'communities' are heterogeneous entities composed by members with distinct capabilities, interests, and demands, as well as characterised by internal conflicts and uneven power relations. Power matters in relationships between the state and indigenous 'communities.' Here, power refers to the relevant political (e.g. knowledge of a system) and social (e.g. friendships, connections, etc.) resources needed to engage in processes of political negotiations with government authorities. Power also matters within indigenous 'communities.' Here, power determines whose interests are taken forward and addressed and whose interests are ignored. This book revealed that it was mainly community leaders, predominantly older men, who played a key role in negotiating rights to the city claims with local authorities. But community leaders rarely had the collective interests and well-being of other community members in mind. Instead, they often (ab)used their position to enrich themselves personally and to preserve their powerful position within the community. In the meantime, the specific interests and demands of women and young people were often not addressed (see Chapter 7).

Indigenous rights to the city struggles, hence, produce winners and losers within indigenous 'communities,' often leading to the marginalisation and exclusion of women and young people. To date, insufficient attention has been paid to such intra-community struggles. Only in La Paz it was possible to identify an organisation – the youth tribe which operates in Chasquipampa and Ovejuyo – which openly confronts the challenge of uneven community representation. What started with personal discussions between a small group of school friends shifted into becoming an alternative urban indigenous organisation where young people, with support from local NGOs, receive organisational training. Through radio shows, rap events, and protest action the youth tribe condemns unjust practices of community leaders. In other words, conflict becomes a central element of relating to other community members and this brings into light existing problems and hierarchies within communities. Understood as such, conflict is an essential component for more emancipatory and inclusive indigenous community relations.

For the members of the indigenous youth tribe, and for other indigenous residents engaged in this research, the struggle for indigenous rights
to the city continues. For this reason, no definitive conclusion is given in
this book. What is evident, however, is that the interests and demands of
different urban indigenous residents, as well as the practices of different
social actors involved in the translation of indigenous rights to the city, may
change over time and vary between different places. Within each place, residents of different ages, genders, and political backgrounds are also aware of
their immediate and long-term challenges, often have ideas in mind on how
to address them in negotiation, contestation, or via confrontational practices. Yet, not everyone manages to meet their interests, needs and rights-
based claims. To succeed in their struggles, these residents, especially those
who are often excluded from decision-making processes within historically
marginalised indigenous communities (e.g. women, young people, etc.), require support from other actors and institutions so that they can voice their
self-determined priorities at different scales. This is likely to be not only
the case for indigenous rights to the city struggles in La Paz and Quito, but
also for the political struggles of other marginalised groups in other parts
of Latin America and the world. Instead of offering general policy recommendations or quick fix solutions, it is therefore best to conclude this book
by calling for more comparative and longitudinal research which captures
the dynamic, conflictive, and constantly shifting nature of right to the city
struggles in different global urban settings. And finally, to ensure that no
one is left behind – a central call highlighted in the New Urban Agenda –
engaged scholars, activists, and practitioners would do well to identify and
strengthen existing emancipatory interventions which not only confront unjust government practices but equally condemn uneven and gendered power
dynamics within historically marginalised communities.

References

Achtenberg, E. 2017. *The Growing Resistance to Megadams in Bolivia.* NACLA Reporting on the Americas.

Albo, X. 1987. 'From MNRistas to Kataristas to Katari.' In *Resistance, Rebellion, and Consciousness in the Andean peasant World Eighteenth to Twentieth Centuries,* S. Stern (ed.), Madison, WI, University of Wisconsin Press, pp. 379–419.

Albo, X. 1991. 'El retorno del indio.' *Revista Andina* 9(2): 299–366.

Albo, X. 2005. *Etnicidad y movimientos indigenas en América Latina.* La Paz, CIPCA.

Albo, X. 2006. 'El Alto, La Vorágine de Una Ciudad Única.' *Journal of Latin American Anthropology* 11(2): 329–350.

Albo, X., Greaves, T. and Sandoval, G. 1981. *Chukiyawu, La cara Aymara de La Paz. El paso a la ciudad.* La Paz, CIPCA.

Albo, X., Greaves, T. and Sandoval, G. 1983. *Chukiyawu, La cara Aymara de La Paz. Cabalgando entre dos mundos.* La Paz, CIPCA.

Anaya, S.J. 2004. *Indigenous Peoples in International Law.* Oxford, Oxford University Press.

Andolina, R., Laurie, N. and Radcliffe, S. 2009. *Indigenous Development in the Andes: Culture, Power, and Transnationalism.* Durham, NC, Durham University Press.

Anthias, P. 2017. *Limits to Decolonization. Indigeneity, Territory and Hydrocarbon Politics in the Bolivian Chaco.* Ithaca, NY, Cornell University Press.

Appadurai, A. 2004. 'The Capacity to Aspire: Culture and the Terms of Recognition.' In *Culture and Public Action,* V. Rao and M. Walton (eds.), Stanford, CA, Stanford University Press, pp. 59–84.

Arbona, J.M. and Kohl, B. 2004. 'City Profile: La Paz – El Alto.' *Cities* 21(3): 255–265.

Assies, W. 1994. 'Self-Determination and the 'New Partnership': The Politics of Indigenous Peoples and States.' In *Indigenous Peoples' Experiences with Self-Government,* W. Assies and A.J. Hoekma (eds.), Copenhagen, IWGIA, pp. 31–71.

Assies, W. 2003. 'David versus Goliath in Cochabamba: Water Rights, Neoliberalism, and the Revival of Social Protest in Bolivia.' *Latin American Perspectives* 30(3): 14–36.

Assies, W. 2010. 'The Limits of State Reform and Multiculturalism in Latin America: Contemporary Illustrations.' In *Out of the Mainstream: Water Rights, Politics and Identity,* R. Boelens, D. Getches and A. Guevara-Gil (eds.), London, Earthscan, pp. 57–73.

Assies, W., Zoomers, A. and Haar, G. 2000. *Land, Territories and Indigenous Peoples' Rights.* Amsterdam, Royal Tropical Institute.

Auyero, J. 2000. *Poor People's Politics: Peronist Survival Networks and the Legacy of Evita*. London, Duke University Press.

Banks, N. 2014. 'Livelihood Limitations: The Political Economy of Urban Poverty in Bangladesh.' *BWPI Working Paper*, Manchester, University of Manchester.

Barsky, O. 1984. *Acumulación Campesina en el Ecuador*. Quito, FLACSO.

Bebbington, A. 1999. 'Capitals and Capabilities: A Framework for Analysing Peasant Viability, Rural Livelihoods and Poverty.' *World Development* 27(12): 2021–2044.

Bebbington, A. 2007. 'Social Movements and the Politicization of Chronic Poverty.' *Development and Change* 38(5): 793–818.

Bebbington, A. and Bebbington, D.H. 2011. 'An Andean Avatar: Post-neoliberal and Neoliberal Strategies for Securing the Unobtainable.' *New Political Economy* 16(1): 131–145.

Becker, M. 2010. *Pachakutik: Indigenous Movements and Electoral Politics in Ecuador*. New York, Rowman & Littlefield Publishers.

Becker, M. 2011. 'Correa, Indigenous Movements, and the Writing of a New Constitution in Ecuador.' *Latin American Perspectives* 38(1): 47–62.

Becker, M. 2013. 'The Stormy Relations between Rafael Correa and Social Movements in Ecuador.' *Latin American Perspectives* 40(3): 43–62.

Bengoa, J. 2000. *La emergencia indigena en America Latina*. Santiago, Fondo de Cultura Economica.

Bourdieu, P. 1977. *Outline of a Theory of Practice*. Cambridge, Cambridge University Press.

Bourdieu, P. 1986. 'The Forms of Capital.' In *Handbook of Theory and Research in Sociology of Education*, J. Richardson (ed.), Westport, CT, Greenwood Press, pp. 241–258.

Bowen, J.D. 2011. 'Multicultural Market Democracy: Elites and Indigenous Movements in Contemporary Ecuador.' *Journal of Latin American Studies* 43(3): 451–483.

Butler, C. 2012. *Henri Lefebvre. Spatial Politics, Everyday Life and the Right to the City*. London, Routledge.

Campbell, S. and Fainstein, S.S. 2003. 'Introduction: The Structure and Debates of Planning Theory.' In *Readings in Planning Theory* (2nd ed.), S. Campbell and S.S. Fainstein (eds.), Oxford, Blackwell, pp. 1–16.

Canessa, A. 2006. 'Todos somos indigenas: Towards a New Language of National Political Identity.' *Bulletin of Latin American Research* 25(2): 241–263.

Canessa, A. 2007. 'Who is Indigenous? Self-Identification, Indigeneity and Claims to Justice in Contemporary Bolivia.' *Urban Anthropology* 36(3): 195–237.

Canessa, A. 2008. 'The Past is Not Another Country: Exploring Indigenous Histories in Bolivia.' *History and Anthropology* 19(4): 353–369.

Canessa, A. 2012. *Intimate Indigeneities: Race, Sex, and History in the Small Spaces of Andean Life*. Durham, NC, Duke University Press.

Cardoso, A.C., Silva, H., Melo, A.C. and Arauja, D. 2018. 'Urban Tropical Forest: Where Nature and Human Settlements Are Assets for Overcoming Dependency, but How Can Urbanisation Theories Identify These Potentials?' In *Emerging Urban Spaces: A Planetary Perspective*, P. Horn, A.C. Cardoso and P. Alfaro d'Alencon (eds.), Cham, Springer, pp. 177–200.

Carter, M. and Barrett, C.B. 2006. 'The Economics of Poverty Traps and Persistent Poverty: An Asset-based Approach.' *Journal of Development Studies* 42(2): 178–199.

Castriota, R. and Tonucci, J. 2018. 'Extended Urbanization in and from Brazil.' *Environment and Planning D: Society and Space* 36(3): 512–528.

Chambers, R. 1994. 'The Origins and Practice of Participatory Rural Appraisal.' *World Development* 22(7): 953–969.

Colloredo-Mansfeld, R. 2009. *Fighting like a Community: Andean Civil Society in an Era of Indian Uprisings.* Chicago, IL, The University of Chicago Press.

Colque, G. 2009. *Municipios de las tierras altas: Breve mapeo para la implementación de las autonomías indígenas.* La Paz, Fundación Tierra.

Coombes, B., Johnson, J.T. and Howitt, R. 2012. 'Indigenous Geographies I: Mere Resource Conflicts? The Complexities in Indigenous Land and Environmental Claims.' *Progress in Human Geography* 36(6): 810–821.

COOTAD. 2010. *Código Orgánico de Organización Territorial, Autonomía y Descentralización.* Quito, Presidencia de la República del Ecuador.

CPE Bolivia. 2009. *Constitución Política del Estado de Bolivia.* La Paz, Vicepresidencia del Estado.

CPE Ecuador. 2008. *Constitución Política del Estado de Ecuador.* Quito, Asamblea Nacional de la Republica del Ecuador.

Crabtree, J. and Chaplin, A. 2013. *Bolivia: Processes of Change.* London, Zed Books.

De Certeau, M. 1984. *The Practice of Everyday Life.* Berkeley, CA, University of California Press.

de Haan, A. 1999. 'Social Exclusion: Towards an Holistic Understanding of Deprivation.' *Social Development Department Dissemination Note No. 2*, London, Department for International Development (DFID).

de la Cadena, M. 2000. *Indigenous Mestizos: The Politics of Race and Culture in Cusco, Peru, 1919–1991.* Durham, NC, Duke University Press.

De Leon, P. and de Leon, L. 2002. 'What Ever Happened to Policy Implementation? An Alternative Approach.' *Journal of Public Administration Research and Theory* 12(4): 467–492.

DeFilippis, J. 2001. 'The Myth of Social Capital in Community Development.' *Housing Policy Debate* 12(4): 781–806.

del Popolo, F., Oyarce, A.M. and Ribotta, B. 2009. 'Indígenas urbanos en América Latina: algunos resultados censales y su relación con los Objetivos de Desarrollo del Milenio' *CEPAL notas de población 86*, Santiago, United Nations Economic Commission for Latin America and the Caribbean.

DMQ, Distrito Metropolitano de Quito. 2003. *Centro Historíco de Quito – Plan Especial.* Quito, Municipio de Quito.

DMQ, Distrito Metropolitano de Quito. 2012a. *Boletin Estatistico Mensual Noviembre.* Quito, Municipio de Quito.

DMQ, Distrito Metropolitano de Quito. 2012b. *Plan Metropolitano de Desarrollo 2012–2022.* Quito, Municipio de Quito.

DMQ, Distrito Metropolitano de Quito. 2012c. *Plan Metropolitano del Ordenamiento Territorial 2012–2022.* Quito, Municipio de Quito.

Dunkerley, J. 2007. *Bolivia: Revolution and the Power of History in the Present.* London, Institute for the Study of the Americas.

Dussel, E. 1993. 'Eurocentrism and Modernity (Introduction to the Frankfurt Lectures).' *Boundary 2* 20(3): 65–76.

Dyck, N. 1985. *Indigenous Peoples and the Nation State, Fourth World Politics in Canada, Australia and Norway.* St Johns, Memorial University of Newfoundland.

Elwood, S., Bond, P. Martinez Novo, C. and Radcliffe, S. 2016. 'Learning from Post-neoliberalisms.' *Progress in Human Geography.* doi:10.1177/0309132516648539.

Engerman, S.L. and Sokoloff, K.L. 2000. 'Factor Endowments, Inequality, and Paths of Development among New World Economies.' *Economia* 3: 41–88.

Escobar, A. 2010. 'Latin America at a Crossroads: Alternative Modernization, Post-liberalism, or Post-development?' *Cultural Studies* 24(1): 1–65.

Espin, M.A. 2012. 'Los indigenas y el espacio citadino. Los lugares de vivienda.' In *San Roque: indígenas urbanos, seguridad y patrimonio*, E. Kingman (ed.), Quito, FLACSO, pp. 101–134.

Espinoza, A.M. 2004. *Viaje a Territorio Projimo.* La Paz, Universidad Mayor de San Andres.

Fenton, S. 2003. *Ethnicity.* Cambridge, Polity Press.

Field, L.W. 1994. 'Review: Who Are the Indians? Reconceptualising Indigenous Identity, Resistance, and the Role of Social Science in Latin America.' *Latin American Research Review* 29(3): 237–248.

Filho, C.M.C. and Goncalvez, R.S. 2010. 'The National Development Plan as a Political Economic Strategy in Evo Morales's Bolivia: Accomplishments and Limitations.' *Latin American Perspectives* 37(4): 177–196.

Flyvbjerg, B. 2003. 'Rationality and Power.' In *Readings in Planning Theory* (2nd ed.), S. Campbell and S.S. Fainstein (eds.), Oxford, Blackwell, pp. 318–331.

Flyvbjerg, B. 2006. 'Five Misunderstandings about Case-study Research.' *Qualitative Inquiry* 12(2): 219–245.

Flyvbjerg, B. 2009. 'Phronetic Planning Research: Theoretical and Methodological Reflections.' *Planning Theory & Practice* 5(3): 283–306.

Fontana, L. 2014. 'The Indigenous Native Peasant Trinity: Imagining a Plurinational Community in Evo Morales's Bolivia.' *Environment and Planning D* 32(3): 1–17.

Gacitúa-Marió, E. and Norton, A. 2009. 'Increasing Social Inclusion through Social Guarantees.' In *Building Equality and Opportunity through Social Guarantees: New Approaches to Public Policy and the Realization of Rights*, E. Gacitúa-Marió, A. Norton and S.V. Georgieva (eds.), Washington, DC, World Bank, pp. 21–32.

Gilbert, A. 2006. 'Good Urban Governance: Evidence from a Model City?' *Bulletin of Latin American Research* 25(3): 392–419.

Gill, L. 2000. *Teetering on the Rim: Global Restructuring, Daily Life, and the Armed Retreat of the Bolivian State.* New York, Colombia University Press.

Ginieniewicz, J. 2015. 'Argentine Migrants to Spain and Returnees: A Case for Accumulation of Civic Assets.' *International Migration* 53(1): 148–170.

Goldstein, D. 2004. *The Spectacular City. Violence and Performance in Urban Bolivia.* Durham, NC, Duke University Press.

Goldstein, D. 2013. *Outlawed: Between Security and Rights in a Bolivian City.* Durham, NC, Duke University Press.

Grugel, J. and Riggirozzi, P. 2012. 'Post-neoliberalism in Latin America: Rebuilding and Reclaiming the State After Crisis.' *Development and Change* 43(1): 1–21.

Gudynas, E. 2011. 'Buen Vivir: Today's Tomorrow.' *Development* 54(4): 441–447.

Guss, D.M. 2006. 'The Gran Poder and the Reconquest of La Paz.' *Journal of Latin American Anthropology* 11(2): 294–328.

Hale, C. 2002. 'Neoliberal Multiculturalism.' *PoLAR: Pollical and Legal Anthropology Review* 28(1): 10–19.

Hall, P. and Tewdwr-Jones, M. 2010. *Urban and Regional Planning.* London, Routledge.

Hardoy, J.E. 1973. *Pre-Colombian Cities.* Toronto, Fitzhenry & Whiteside.

Hardoy, J.E. 1989. 'The Legal and the Illegal City.' In *Squatter Citizen: Life in the Urban Third World,* J.E. Hardoy and D. Satterthwaite (eds.), London, Earthscan, pp. 12–36.

Harris, O. 1995. 'Identity and Market Relations: Indians and Mestizos in the Andes.' In *Ethnicity, Markets and Migration in the Andes,* O. Harris, B. Larson and E. Tandeter (eds.), Durham, NC, Duke University Press, pp. 351–390.

Harris, O. and Tandeter, E. 1987. *La participación indígena en los mercados surandinos: Estrategias y reproducción social siglos XVI a XX.* La Paz, Centro de Estudios de la Realidad Económica y Social.

Harvey, D. 2008. 'The Right to the City.' *New Left Review* 53: 23–40.

Harvey, D. 2012. *Rebel Cities.* London, Verso.

Hay, C. 2002. *Political Analysis.* Basingstoke, Palgrave.

Healey, P. 1997. *Collaborative Planning: Shaping Places in Fragmented Societies.* Vancouver, UBC Press.

Hickey, S. and du Toit, A. 2007. 'Adverse-Incorporation, Social Exclusion and Chronic Poverty.' *CPRC Working Paper 81,* Manchester, Chronic Poverty Research Centre.

Hickey, S. and Mitlin, D. 2009. 'Introduction.' In *Rights-based Approaches to Development: Exploring the Potential and Pitfall,* D. Mitlin and S. Hickey (eds.), Sterling, VA, Kumarian Press, pp. 3–20.

Hope, J. 2016. 'Losing Ground? Extractive-led Development versus Environmentalism in the Isiboro Secure Indigenous Territory and National Park (TIPNIS), Bolivia.' *The Extractive Industries and Society* 3(4): 922–929.

Horn, P. 2018. 'Emerging Urban Indigenous Spaces in Bolivia: A Combined Planetary and Postcolonial Perspective.' In *Emerging Urban Spaces: A Planetary Perspective,* P. Horn, A.C. Cardoso and P. Alfaro d'Alencon (eds.), Cham, Springer, pp. 43–64.

Horn, P. and Grugel, J. 2018. 'The SDGs in Middle-Income Countries: Setting or Serving Domestic Development Agendas? Evidence from Ecuador.' *World Development* 109: 73–84.

Horn, P., Mitlin, D., Bennett, J., Chitekwe-Biti, B. and Makau, J. 2018. 'Towards Citywide Participatory Planning: Emerging Community-led Practices in Three African Cities.' *GDI Working Paper,* Manchester, The University of Manchester.

Hornberger, N.H. 2000. 'Bilingual Education Policy and Practice in the Andes: Ideological Paradox and Intercultural Possibility.' *Anthropology & Education Quarterly* 31(2): 173–201.

Hornberger, N.H. and Swijnhart, K.F. 2013. 'Bilingual Intercultural Education and Andean Hip Hop: Transnational Sites for Indigenous Language and Identity.' *Language in Society* 41: 499–525.

Imilan, W.A. 2010. *Warriache-Urban Indigenous: Mapuche Migration and Ethnicity in Santiago de Chile.* Munster, LIT Verlag.

INE. 2014. *Resultados: Censo de Poblacion y Vivienda 2012.* La Paz, Instituto Nacional de Estadística.

INEC. 2014. *Censo de Poblacion y Vivienda 2010.* Quito, Instituto Nacional de estadística y censos.

Jojola, T. 2008. 'Indigenous Planning: An Emerging Context.' *Canadian Journal of Urban Research* 17(1): 37–47.

Keith, R.G. 1971. 'Encomienda, Hacienda, and Corregimiento in Spanish America: A Structural Analysis.' *Hispanic American Historical Review* 52(3): 431–446.

Kingman, E. 2012. 'Ciudad, seguridad y racismo.' In *San Roque: indígenas urbanos, seguridad y patrimonio*, E. Kingman (ed.), Quito, FLACSO, pp. 175–209.

Klein, H.S. 2011. *A Concise History of Bolivia* (2nd ed.). New York, Cambridge University Press.

Klein, H.S. and Vinson, B. 2006. *African Slavery in Latin America and the Caribbean*. Oxford, Oxford University Press.

Klor de Alva, J. 1992. 'Colonialism and Postcolonialism as (Latin) American Mirages.' *Colonial Latin American Review* 1(1): 3–23.

Kohl, B. and Farthing, L. 2006. *Impasse in Bolivia*. London, Zed Books.

Korovkin, T. 1992. 'Indians, Peasants, and the State: The Growth of a Community Movement in the Ecuadorian Andes.' *CERLAC Occasional Paper Series*, Ontario, York University.

Kothari, U. 2005. *A Radical History of Development Studies: Individuals, Institutions and Ideologies*. London, Zed Books.

Kymlicka, W. 1995. *Multicultural Citizenship: A Liberal Theory of Minority Rights*. Oxford, Oxford University Press.

LAD. 2010. *Ley marco de autonomias y descentralizacion 'Andres Ibañez'*. La Paz, Ministerio de Planificación del Desarrollo.

Landau, D. 2017. 'Presidential Term Limits in Latin America and the Limits of Transnational Constitutional Dialogue.' *Public Law Research Paper 862*, Tallahassee, FL, Florida State University College of Law.

La Paz. 1995. *Ley Numero 1669*. La Paz, Gobierno Municipal.

La Paz. 2006. *Atlas del municipio de La Paz*. La Paz, Gobierno Municipal.

La Paz. 2010. *La Paz 10 años en cifras 2000–2009. Compendia Estadistico de Bicentario*. La Paz, Gobierno Municipal.

La Paz. 2012. *Proyecto Carta Orgánica del Municipio de Nuestra Señora de La Paz*. La Paz, Gobierno Municipal.

Larson, B. 2004. *Trials of Nation Making: Liberalism, Race and Ethnicity in the Andes: 1810–1910*. Cambridge, Cambridge University Press.

Lazar, S. 2004. 'Personalist Politics, Clientelism and Citizenship: Local Elections in El Alto, Bolivia.' *Bulletin of Latin American Research* 23(2): 228–243.

Lazar, S. 2008. *El Alto, Rebel City. Self and Citizenship in Andean Bolivia*. Durham, NC, Duke University Press.

LCP. 2010. *Ley Orgánica de Participación Ciudadana*. Quito, Presidencia de la República del Ecuador.

LE. 2010. *Ley de Educación Avelino Siñani-Elizardo Pérez*. La Paz, Ministerio de Educación.

Lefebvre, H. 1968. *Le droit à la ville*. Paris, Editions Anthropos.

Lefebvre, H. 1991. *The Production of Space*. Oxford, Blackwell.

LIE. 2011. *Ley Orgánica de Educación Intercultural*. Quito, Presidencia de la República del Ecuador.

Lipsky, M. 1980. *Street-Level Bureaucracy*. New York, Russell Sage.

LJD. 2010. *Ley de deslinde jurisdiccional*. La Paz, Ministerio de Justicia.

Lozano, A. 1991. *Quito: Ciudad Milenaria*. Quito, Ediciones Abya Yala.

LPP. 1994. *Ley de participación popular*. La Paz, Lexivox.

LPS. 2013. *Ley de participación y control social*. La Paz, Ministerio de transparencia institucional y lucha contra la corrupcion.

LRPRUEH. 2012. *Ley de regularización del derecho propietario sobre bienes inmuebles urbanos destinados a vivienda*. La Paz, Ministerio de Planificación del Desarrollo.

Lucero, J.A. 2004. 'Indigenous Political Voice and the Struggle for Recognition in Ecuador and Bolivia.' In *Institutional Pathways to Equity: Addressing Inequality Traps*, A. Bebbington, A.A. Dani, A. de Haan and M. Walton (eds.), Washington, DC, World Bank, pp. 139–168.

Maclean, K. 2018. 'Envisioning Gender, Indigeneity and Urban Change: The Case of La Paz, Bolivia.' *Gender, Place & Culture: A Journal of Feminist Geography* 25(5): 711–726.

Mahoney, J. 2003. 'Long-Run Development and the Legacy of Colonialism in Spanish America.' *The American Journal of Sociology* 109(1): 50–106.

Marcuse, P. 2009. 'From Critical Theory to the Right to the City.' *City* 13(2–3): 185–197.

Marti i Puig, S. 2010. 'The Emergence of Indigenous Movements in Latin America and their Impact on the Latin American Political Scene: Interpretative Tools at the Local and Global Levels.' *Latin American Perspectives* 37(6): 74–92.

Martínez Cobo, J. 1987. *Study of the Problem of Discrimination against Indigenous Populations, Volume V, Conclusions, Proposals and Recommendations*. New York, United Nations.

Martinez Novo, C. 2014. 'The Minimization of Indigenous Numbers and the Fragmentation of Civil Society in the 2010 Census in Ecuador.' *Journal of Iberian and Latin American Research* 20(3): 399–422.

Matos Mar, J. 1957. *Las Barriadas de Lima*. Lima, Instituto de Estudios Peruanos.

McIlwaine, C. 2011. 'Constructing Transnational Social Spaces among Latin American Migrants in Europe: Perspectives from the UK.' *Cambridge Journal of Regions, Economy and Society* 5(2): 289–304.

McNeish, J.A. 2013. 'Extraction, Protest and Indigeneity in Bolivia: The TIPNIS Effect.' *Latin American and Caribbean Ethnic Studies* 8(2): 221–242.

Merrifield, A. 2011. 'The Right to the City and Beyond: Notes on a Levebvrian re-Conceptualisation.' *City* 15(3–4): 468–476.

Mignolo, W. 2000. *Local Histories/Global Designs: Coloniality, Subaltern Knowledges, and Border Thinking*. Princeton, NJ, Princeton University Press.

Milton, C.E. 2005. 'Poverty and the Politics of Colonialism: "Poor Spaniards," Their Petitions, and the Erosion of Privilege in Late Colonial Quito.' *Hispanic American Historical Review* 85(4): 595–626.

Miraftab, F. 2009. 'Insurgent Planning: Situating Radical Planning in the Global South.' *Planning Theory* 8(1): 32–50.

Mitlin, D. and Satterthwaite, D. 2013. *Urban Poverty in the Global South. Scale and Nature*. London, Routledge.

Molyneux, M. and Lazar, S. 2003. *Doing the Rights Thing. Rights-Based Development and Latin American NGOs*. London, ITDG Publishing.

Monte-Mor, R.L. 2018. 'Urbanisation, Sustainability and Development: Contemporary Complexities and Diversities in the Production of Urban Space.' In *Emerging Urban Spaces: A Planetary Perspective*, P. Horn, A.C. Cardoso and P. Alfaro d'Alencon (eds.), Cham, Springer, pp. 201–216.

Morse, R.M. 1978. 'Latin American Intellectuals and the City, 1860–1940.' *Journal of Latin American Studies* 10(2): 219–238.

Moser, C.O.N. 1993. *Gender Planning and Development*. London, Routledge.

Moser, C.O.N. 1998. 'The Asset Vulnerability Framework: Reassessing Urban Poverty Reduction Strategies.' *World Development* 26(1): 1–19.

Moser, C.O.N. 2009. *Ordinary Families, Extraordinary Lives: Assets and Poverty Reduction in Guayaquil, 1978–2004*. Washington, DC, The Brookings Institution.

Moser, C.O.N. and Horn, P. 2015. 'Does Economic Crisis Always Harm International Migration? Longitudinal Evidence from Ecuadorians in Barcelona.' *International Migration* 53(2): 274–290.

Myers, G.A. 2003. *Verandahs of Power: Colonialism and Space in Urban Africa*. Syracuse, NY, Syracuse University Press.

Oehmichen, C. 2001. 'Espacio urbano y segregación étnica en la ciudad de México.' *Population Papers 28*, Mexico D.F., Universidad Autónoma del Estado de México.

Perreault, T. 2006. 'From the *Guerra del Agua* to the *Guerra del Gas:* Resource Governance, Neoliberalism and Popular Protest in Bolivia.' *Antipode* 38(1): 150–172.

Pierre, J. 1999. 'Models of Urban Governance: The Institutional Dimension of Urban Politics.' *Urban Affairs Review* 34(3): 372–396.

Platt, T. 1982. *Estado boliviano y ayllu andino: tierra y tributo en el Norte de Potosi*. Lima, Instituto de Estudios Peruanos.

Porter, L. 2010. *Unlearning the Colonial Cultures of Planning*. London, Routledge.

Porter, L. and Barry, J. 2016. *Planning for Coexistence? Recognizing Indigenous Rights Through Land-use Planning in Canada and Australia*. London, Routledge.

Postero, N.G. 2007. *Now We Are Citizens: Indigenous Politics in Postmulticultural Bolivia*. Stanford, CA, Stanford University Press.

Postero, N. 2013. 'Introduction: Negotiating Indigeneities.' *Latin American and Caribbean Ethnic Studies* 8(2): 107–121.

Postero, N. 2017. *The Indigenous State: Race, Politics, and Performance in Plurinational Bolivia*. Oakland, CA, University of California Press.

Prusak, S.Y., Walker, R. and Innes, R. 2016. 'Toward Indigenous Planning? First Nation Community Planning in Saskatchewan, Canada.' *Journal of Planning Education and Research* 36(4): 440–450.

Purcell, M. 2002. 'Excavating Lefebvre: The Right to the City and Tts Urban Politics of the Inhabitant.' *GeoJournal* 58: 99–108.

Putnam, R. 1993. Making Democracy work: civic traditions in modern Italy. Princeton, Princeton University Press.

Quijano, A. 1975. 'The Urbanization of Latin American Society.' In *Urbanization in Latin America: Approaches and Issues*, J. Hardoy (ed.), New York, Anchor Books.

Quijano, A. 1993. 'Raza, etnia y nación: Cuestiones Abiertas.' *Estudios Latinoamericanos* 2(3): 757–775.

Quijano, A. 2000. 'Coloniality of Power and Eurocentrism in Latin America.' *International Sociology* 15(2): 215–232.

Quijano, A. 2005. 'Of Don Quixote and Windmills in Latin America.' *Estudos Avançados* 19(55): 2–22.

Quijano, A. 2006. 'El "movimiento indígena" y las cuestiones pendientes en América Latina.' *Argumentos* 19(50): 51–77.

Quijano, A. 2007. 'Coloniality and Modernity/Rationality.' *Cultural Studies* 21(2): 168–178.

Rayner, J. 2017. 'The Struggle for Quito's Communes: Negotiating Property and Citizenship in Plurinational, Post-neoliberal Ecuador.' *Urban Anthropology* 46(1): 95–134.

Reinega, F. 1970. *La Revolucion India*. La Paz, Minka.

Revilla, C. 2011. 'Understandings the Mobilizations of Octubre 2003: Dynamic Pressures and Shifting Leadership Practices in El Alto.' In *Remapping Bolivia: Resources, Territory, and Indigeneity in a Plurinational State*, N. Fabricant and B. Gustafson (eds.), Santa Fe, NM, School of American Research Press, pp. 121–145.

Risor, H. 2010. 'Twenty Hanging Dolls and a Lynching: Defacing Dangerousness and Enacting Citizenship in El Alto, Bolivia.' *Public Culture* 22(3): 465–485.

Rivera Cusicanqui, S. 2010. 'The Notion of "Rights" and the Paradoxes of Postcolonial Modernity: Indigenous Peoples and Women in Bolivia.' *Qui Parle: Critical Humanities and Social Sciences* 18(2): 29–54.

Robins, S., Cornwall, A. and Von Lieres, B. 2008. 'Rethinking 'Citizenship' in the Postcolony.' *Third World Quarterly* 29(6): 1069–1086.

Rojo, C.C. 2012. 'Construccion de idendidades de las vendedoras Kichwas y mestizas y los juegos de poder en el Mercado San Roque.' In *San Roque: indígenas urbanos, seguridad y patrimonio*, E. Kingman (ed.), Quito, FLACSO, pp. 79–100.

Roy, A. 2009. 'Why India Cannot Plan its Cities: Informality, Insurgence and the Idiom of Urbanization.' *Planning Theory* 8(1): 76–87.

Russel, P.H. 2005. *Recognizing Aboriginal Title: The Mabo Case and Indigenous Resistance to English Settler-Colonialism*. Toronto, University of Toronto Press.

Salmoral, M.L. 1994. 'La ciudad de Quito hacia mil ochocientos.' *Anuario de Estudios Americanos* 51(1): 143–64.

Salomon, F. 1988. 'Indian Women of Early Colonial Quito as Seen Through their Testaments.' *The Americas* 44(3): 325–341.

Sandercock, L. 2003. *Towards Cosmopolis: Planning for Multicultural Cities*. London, Wiley.

Satterthwaite, D. 2008. 'Building Homes: The Role of Federations of the Urban Poor.' In *Assets, Livelihoods, and Social Policy*, C. Moser and A. Dani (eds.), Washington, DC, World Bank, pp. 171–194.

Schavelzon, S. 2013. *El nacimiento del Estado Plurinacional de Bolivia*. La Paz, Plural Editores.

Scheman, L.R. 1988. *The Alliance for Progress: A Retrospective*. London, Praeger.

Scott, J.C. 1999. *Seeing like a State. How Certain Schemes to Improve the Human Condition Have Failed*. New Haven, Yale University Press.

Sen, A. 1981. *Poverty and Famines: An Essay on Entitlement and Deprivation*. Oxford, Oxford University Press.

Sen, A. 1997. 'Editorial: Human Capital and Human Capability.' *World Development* 25(12): 1959–1961.

SENPLADES. 2009. *Plan Nacional para el Buen Vivir 2009*. Quito, Secretaria Nacional de Planificación y Desarrollo.

SENPLADES. 2013. *Good Living National Plan 2013–2017: A Better World for Everyone*. Quito, Secretaria Nacional de Planificación y Desarrollo. Sherraden, M. 1991. *Assets and the Poor: A New American Welfare Policy*. Armok, NY, M.E. Sharpe.

Sieder, R. 2002. 'Introduction.' In *Multiculturalism in Latin America: Indigenous Rights, Diversity, and Democracy*, R. Sieder (ed.), Houndmills, Palgrave Macmillan, pp. 1–23. Smith, L.T. 1999. *Decolonising Methodologies: Research and Indigenous Peoples*. New York, St Martin's Press.

Sousz, P., Yampara, S., Zaratti, A., Medina, J., Saavedra, J.L., Portugal, P., Cuadros, D. and Gallardo, H. 2010. *Matrices Civilizatorias: Construcción de Políticas Municipales Interculturales*. La Paz, Oxfam.

Speiser, S. 2004. *Indigene Völker in Städten: präsent und doch nicht wahrgenommen.* Eschborn, GIZ.Starn, O. 1991. 'Missing the Revolution: Anthropologists and the War in Peru.' *Cultural Anthropology* 6(1): 63–69.

Stavenhagen, R. 1981. *Siete tesis equivocadas sobre América.* México D.F., Nuestro Tiempo.

Stefanoni, P. 2012. 'Posneoliberalismo custa arriba: Los modelos de Venezuela, Bolivia y Ecuador en debate.' *Nueva Sociedad* 239: 51–64.

Stein, A. and Horn, P. 2012. 'Asset Accumulation: An Alternative to Achieving the Millennium Development Goals.' *Development Policy Review* 30(6): 663–680.

Swanson, C. 2007. 'Revanchist Urbanism Heads South: The Regulation of Indigenous Beggars and Street Vendors in Ecuador.' *Antipode* 39(4): 708–728.

Tassi, N. 2010. 'The Postulate of Abundance. Cholo Market and Religion in La Paz, Bolivia.' *Social Anthropology* 18(2): 191–209.

Tassi, N., Medeiros, C., Rodriguez-Carmona, A. and Ferrufino, G. 2013. 'Hacer plata sin plata.' *Nueva Sociedad* 241: 93–105.

Tilly, C. 1984. *Big Structures, Large Processes, Huge Comparisons.* New York, Russell Sage.

Tockman, J. and Cameron, J. 2014. 'Indigenous Autonomy and the Contradictions of Plurinationalism in Bolivia.' *Latin American Politics and Society* 56(3): 46–69.

Torrico Foronda, E. 2011. 'El Nuevo Rostro Urbano de Bolivia.' In *Ciudades en transformacion. Disputas por el espacio, apropriacion de la ciudad y practicas de ciudadania*, P. Urquieta (ed.), La Paz, Oxfam, pp. 61–72.

Touraine, A. 2000. 'A Method for Studying Social Actors.' *Journal of World Systems Research* 6(3): 900–918.

Turner, J.F.C. 1968. 'Housing Priorities, Settlement Patterns and Urban Development in Modernizing Countries.' *Journal of the American Institute of Planners* 34(6): 354–363.

Turner, J.F.C. 1976. *Housing by People – Towards Autonomy in Building Environments.* London, Marion Boyars.

Turner, J.F.C. 1978. 'Housing in Three Dimensions: Terms of Reference for the Housing Question Redefined.' *World Development* 6(9/10): 1135–1145.

UN Habitat. 2010. *Urban Indigenous Peoples and Migration: A Review of Policies, Programmes and Practices.* Nairobi, United Nations Human Settlements Programme.

van Cott, D.L. 2000. *The Friendly Liquidation of the Past: The Politics of Diversity in Latin America.* Pittsburgh, University of Pittsburgh Press.

van Cott, D.L. 2008. *Radical Democracy in the Andes.* Cambridge, Cambridge University Press.

van den Berghe, P.L. 1974. 'Introduction.' In *Class and Ethnicity in Peru*, P.L. van den Berghe (ed.), Leiden, Brill, pp. 1–11.

Wade, P. 2010. *Race and Ethnicity in Latin America* (2nd ed.). London, Pluto Press.

Walsh, C. 2010. '"Raza", mestizaje y poder: horizontes coloniales pasados y presentes.' *Crítica y Emancipación* 2(3): 95–126.

Walsh, C. 2011. 'Development as *Buen Vivir*: Institutional Arrangements and (de) colonial Entanglements.' *Development* 53(1): 15–21.

Warren, S.D. 2017. 'Indigenous in the City: The Politics of Urban Mapuche Identity in Chile.' *Ethnic and Racial Studies* 40(4): 694–712.

Weisbrot, M., Johnston, J. and Merling, L. 2017. *Decade of Reform: Ecuador's Macroeconomic Policies, Institutional Changes, and Results.* Washington, DC, Center for Economic and Policy Research.

Weismantel, M. 2001. *Cholos and Pishtacos: Stories of Race and Sex in the Andes.* Chicago, IL, The University of Chicago Press.

Wilson, J. and Bayón, M. 2015. 'Concrete Jungle: The Planetary Urbanization of the Ecuadorian Amazon.' *Human Geography* 8(3): 1–23.

Wilson, S. 2009. *Research is Ceremony. Indigenous Research Methods.* Black Point, Fernwood Publishing.

World Vision. 2012. *Informe del Estudio de Diagnostico para el PDA Palca.* La Paz, World Vision.

Yashar, D. 2005. *Contesting Citizenship in Latin America: The Rise of Indigenous Movements and the Postliberal Challenge.* Cambridge, Cambridge University Press.

Yiftachel, O. 2006. *Ethnocracy: Land and Identity Politics in Israel/Palestine.* Philadelphia, PA, University of Pennsylvania Press.

Yin, R.K. 2003. *Case Study Research: Design and Methods* (3rd ed.). London, Sage Publications.

Zaaijer, M. 1991. 'City Profile: Quito.' *Cities* 8(2): 87–92.

Zibechi, R. 2010. *Dispersing Power: Social Movements as Anti-State Forces.* Edinburgh, AK Press.

Index

Note: Boldface page numbers refer to tables; italic page numbers refer to figures and page numbers followed by "n" denote endnotes.

Printed and bound by CPI Group (UK) Ltd, Croydon, CR0 4YY

24/10/2024

01778279-0020